Diss. ETH No. 16937

Spectral Efficient Cooperative Relaying Strategies for Wireless Networks

A dissertation submitted to the
ETH ZURICH

for the degree of
Doctor of Sciences

presented by
BORIS RANKOV
Dipl. El. Ing. ETH Zurich
born April 23, 1974
citizen of Winterthur (ZH), Switzerland

accepted on the recommendation of
Prof. Dr. Armin Wittneben, examiner
Prof. Dr. Gerhard Fettweis, co-examiner

2007

Reihe Series in Wireless Communications
herausgegeben von:
Prof. Dr. Armin Wittneben
Eidgenössische Technische Hochschule
Institut für Kommunikationstechnik
Sternwartstr. 7
CH-8092 Zürich

E-Mail: wittneben@nari.ee.ethz.ch
Url: http://www.nari.ee.ethz.ch/

Bibliografische Information der Deutschen Nationalbibliothek

Die Deutsche Nationalbibliothek verzeichnet diese Publikation in der
Deutschen Nationalbibliografie; detaillierte bibliografische Daten sind
im Internet über http://dnb.d-nb.de abrufbar.

ISBN 978-3-8325-1744-1
ISSN 1611-2970

Logos Verlag Berlin GmbH
Comeniushof, Gubener Str. 47,
10243 Berlin
Tel.: +49 030 42 85 10 90
Fax: +49 030 42 85 10 92
INTERNET: http://www.logos-verlag.de

Für David und Susanna

Abstract

In this work we are interested in spectral efficient *physical layer cooperation* strategies for wireless networks. We propose and investigate communication protocols that achieve higher spectral efficiencies than existing cooperative protocols. Similar as in multiple-input multiple-output (MIMO) point-to-point systems where the increase in spectral efficiency is due to multiple antennas at the transmitter and the receiver, we want to exploit the availability of multiple antennas in a wireless network. However, since the antennas are located at different geographic locations, some kind of cooperation has to be introduced. By cooperation we mean that distant terminals share their antenna resources, i.e., terminals help other terminals in the communication process. A very important form of cooperation is relaying, where a third terminal relays data from one terminal (source) to another terminal (destination). The capacity of the canonical three-node relay channel is still not found, but tight upper and lower bounds are known. We will see that different relaying strategies lead to different MIMO gains, such as diversity, array (increase of average signal-to-noise ratio) or multiplexing gain [1]. For example, the *decode-and-forward* scheme provides a transmit array gain and/or diversity gain (source and relay transmit the same symbol). The *compress-and-forward* strategy provides a receive array gain and/or diversity gain (relay and destination observe two independent observations of the same source symbol).

In this work we extend the classical relaying protocols to *two-way relay channels* and provide the corresponding achievable rate regions. We show that in an AWGN channel a combined strategy of two-way compress/decode-and-forward achieves the cut-set upper bound on the sum-capacity of the two-way relay channel, when the relay is near to one of the terminals. In a fading two-way relay channel the cut-set upper bound can be achieved by a two-way compress-and-forward strategy.

An often assumed constraint in wireless relay channels is that relays can only operate as half-duplex devices, i.e., relays cannot transmit and receive simultaneously in the same frequency band. Two channel uses are necessary to transmit the data symbols from a source to a destination via a relay. This leads to a factor one-half in front of the log in the capacity expressions for the different relaying strategies. We propose *two-way relaying* and *two-path relaying* in order to increase the pre-log factor and therefore the spectral efficiency. In two-path relaying two relays alternate in transmission and reception. All nodes operate in the same physical channel, thereby causing inter-relay interference. For amplify-and-forward (AF) relays, it is shown that this protocol can recover a significant portion of the half-duplex loss (pre-log factor $\frac{1}{2}$) for weak to moderately strong inter-relay channels. For decode-and-forward (DF) relays we propose three strategies where the half-duplex loss becomes negligible when the inter-relay channel is very weak or very strong and performs reasonably well for moderate inter-relay channels. In two-way relaying, we propose a relaying protocol where a synchronous bidirectional connection between two half-duplex terminals is established using one half-duplex relay. Hereby, the achievable rate in one direction suffers still from the pre-log factor $\frac{1}{2}$ but since two connections are realized in the same physical channel the sum-rate is considerably higher. We show that for DF relays the cut-set upper bound is achieved, when the relay is in the proximity of one of the terminals.

Two applications of relay channels are studied in this thesis. We propose to use relays for spectral efficient signaling in rank-deficient MIMO channels. These channels fail to provide significant multiplexing gains, since there are not enough spatial degrees of freedom to open up parallel spatial channels. We propose to use half-duplex amplify-and-forward relays as active scatterer in a way that the eigenvalue structure of the compound two-hop matrix channel improves with respect to the multiplexing capability. The resulting factor one-half loss in spectral efficiency due to the half-duplex operation of the relays can be circumvented by the spectral efficient relaying protocols proposed in this thesis.

The second application focuses on multi-hop relay systems. We propose two systems, that are capable to achieve a distributed MIMO multiplexing gain with single-antenna nodes. The first system uses AF relays and it is shown that the achievable rate decreases with increasing number of hops due to two effects: accumulation of relay noise along the multi-hop chain and singularity of the product channel matrix. The second system uses DF relays, where it is shown that for a weakest link signaling protocol, the achievable rate also decreases with increasing number of hops, although there is no noise accumulation along the multi-hop chain. For an ergodic signaling protocol the achievable rate is independent of the number of hops.

Kurzfassung

Diese Arbeit befasst sich mit spektral effizienten Kommunikationsprotokollen für drahtlose Netzwerke. Aus der Forschung über Punkt-zu-Punkt Systeme mit Mehrfachantennen an Sender und Empfänger weiss man, dass ein beträchtlicher Gewinn an spektraler Effizienz erreicht werden kann durch geschicktes Ausnutzen der zusätzlichen räumlichen Freiheitsgrade. In dieser Arbeit entwickeln und untersuchen wir Protokolle und Kommunikationsstrategien, die das Vorhandensein von verteilten Antennen in einem drahtlosen Netzwerk ausnützen. Da die Antennen zu verschiedenen Teilnehmern im Netzwerk gehören und geographisch separiert sind, ist eine Kooperation zwischen den Teilnehmern nötig. Eine wichtige Form von Kooperation ist *Relaying*, wobei das *Relay* Datenpakete von einer Quelle empfängt, eine Verarbeitung durchführt und dann das Datenpaket an die Senke oder das nächste Relay weiterleitet. Der Zielempfänger ist dabei häufig auch in der Lage, Datenpakete sowohl direkt vom Sender als auch vom Relay zu empfangen. Die Kapazität (maximale Datenrate) dieses Kanals ist bis heute ein offenes Problem der Informationstheorie.

Wir werden unter anderem sehen, dass mit verschiedenen Relayprotokollen ähnliche Performanceverbesserungen erreicht werden können wie bei MIMO Systemen, wie Diversität, Signal-zu-Geräusch (SNR) Anhebung aufgrund eines Antennenarrays am Sender (*transmit array gain*), SNR Anhebung aufgrund eines Antennenarrays am Empfänger (*receive array gain*) oder räumliche Multiplexierung [1]. Zu den bekanntesten Protokollen zählen *Amplify-and-forward* (AF), *Decode-and-forward* (DF) und *Compress-and-forward* (CF). Beim ersten Protokoll führt das Relay eine einfache lineare Verstärkung durch, bevor das Signal weitergeleitet wird. Das zweite Protokoll basiert darauf, dass das Relay das gesamte Datenpaket dekodiert und danach nochmals kodiert, bevor es weitergesendet wird. Beim dritten Protokoll wird das Empfangssignal am Relay quantisiert, komprimiert und danach weitergeleitet. Jedes Verfahren hat seine Vor- und Nachteile, und je nach Situation ist die spektrale Effizienz unterschiedlich.

In dieser Arbeit erweitern wir die klassischen Relayprotokolle auf den Fall von bidirektionalem Relaying und leiten die simultan (in zwei Richtungen) erreichbaren Datenraten her. Wir zeigen, dass in einem AWGN Kanal ein kombiniertes Verfahren von CF/DF eine obere Schranke der

Summenkapazität des bidirektionalen Kanals erreichen kann (und in diesem Sinne optimal ist), wenn das Relay in der Nähe der Quelle oder der Senke ist. In einem Schwundkanal hingegen, wird die obere Schranke durch ein bidirektionales CF erreicht.

Eine oft gemachte Annahme in drahtlosen Netzwerken ist, dass das Relay auf einer festen Frequenz nicht gleichzeitig senden und empfangen kann (halb-duplex). Demnach sind zwei Kanalbenutzungen nötig, um ein Symbol von der Quelle zur Senke via das Relay zu übertragen. Dies führt aber zu einem Verlust an spektraler Effizienz, da die möglichen Raten halbiert werden.

Wir schlagen zwei Protokolle vor, *Bidirektionales* und *Zwei-Wege Relaying* vor, welche die spektrale Effizienz massiv verbessern. Beim Zwei-Wege Verfahren nehmen wir zwei halb-duplex Relays an, die abwechselnd senden und empfangen. Dadurch ist die Quelle in der Lage, in jedem Zeitschlitz neue Information zu senden. Die Rate wird nicht mehr halbiert, jedoch entsteht zwischen den Relays Interferenz, welche die Rate etwas reduziert. Wir zeigen, dass mit dem Zwei-Wege Verfahren sowohl für AF als auch DF die spektrale Effizienz markant erhöht werden kann. Im bidirektionalen Verfahren nehmen wir an, dass zwei Teilnehmer gleichzeitig miteinander über ein halb-duplex Relay kommunizieren. Dabei wird wieder die Rate jedes Datenstroms halbiert, aber da zwei Datenströme gleichzeitig über den gleichen physikalischen Kanal übertragen werden, ist die Summenrate höher und die spektrale Effizienz des Netzwerkes deutlich verbessert.

Wir schlagen dann zwei Anwendungen von Relaying vor. Im ersten Beispiel benutzen wir AF Relays, um in schlecht konditionierten MIMO Systemen den Rang und damit die spektrale Effizienz vom MIMO Kanälen zu verbessern. Wenn ein MIMO Kanal zu wenig Streuer aufweist, zeigt sich, dass die Kanalmatrix einen tiefen Rang besitzt und somit nur wenige räumliche Unterkanäle für die Kommunikation geöffnet werden können. In solchen Kanälen setzen wir AF Relays ein, die wie aktive Streuer wirken und wir zeigen, dass der effektive zwei-hop MIMO Kanal einen Ranggewinn erfährt und somit die spektrale Effizienz verbessert wird.

Im zweiten Anwendungsbeispiel zeigen wir, wie in einem Netzwerk bestehend aus Teilnehmern mit jeweils nur einer Antenne ein *multi-hop* MIMO System errichtet werden kann. Wir zeigen, dass für den Fall von AF Relays zwei Effekte die erreichbare Rate negativ beeinflussen: die Akkumulation des Rauschens von den Relays an der Senke und die zunehmende Singularität der zusammengesetzten Kanalmatrix mit zunehmender Anzahl *Hops*. Für den Fall von DF Relays schlagen wir zwei Protokolle für die Signalisierung vor: *Weakest Link Signaling* und *Ergodic Signaling*. Wir zeigen, dass beim ersten Protokoll die Rate abnimmt mit zunehmender Anzahl Hops, wohingegen das zweite Protokoll unabhängig von der Anzahl Hops ist.

Contents

1 Introduction   **11**

  1.1   Motivation 11

  1.2   General Overview of Related Work 13

  1.3   Main Contributions of the Thesis 20

  1.4   Organization of the Thesis 22

2 Information-theoretic Preliminaries   **25**

  2.1   Theory of Typical Sequences 25

      2.1.1   Weak vs. Strong Typicality 25

      2.1.2   Definition and Properties of Strongly Typical Sequences 26

      2.1.3   Jointly and Conditionally Typical Sequences 29

  2.2   Lossless Source Coding 34

      2.2.1   Source Coding Based on AEP 35

      2.2.2   Source Coding Based on Binning 35

  2.3   Lossy Source Coding or Rate Distortion Theory 36

  2.4   Lossy Source Coding with Side Information or The Wyner-Ziv Problem 41

3 Capacity Theorems for the Relay Channel   **47**

  3.1   Capacity Theorems for the Relay Channel 47

      3.1.1   System Model 47

      3.1.2   Cut-set Upper Bound 49

      3.1.3   Decode-and-forward 51

      3.1.4   Compress-and-forward 57

      3.1.5   Amplify-and-forward 64

      3.1.6   Numerical Examples 66

  3.2   Application to Fading Relay Channels 70

      3.2.1   System Model 70

3.2.2   Wireless Full-duplex Relaying 71

3.2.3   Wireless Half-duplex Relaying 73

3.2.4   Numerical Examples 78

4   Achievable Rate Regions for the Two-way Relay Channel   85

4.1   System Model 86

4.2   Two-way Decode-and-forward 88

4.3   Two-way Compress-and-forward 93

4.4   Amplify-and-forward 102

4.5   Combined Decode/Compress-and-Forward 103

4.6   Numerical Examples 106

5   Spectral Efficient Protocols for Two-hop Half-Duplex Relay Channels   109

5.1   Spectral Efficiency of Two-hop Half-Duplex Relaying Protocols 111

     5.1.1   Amplify-and-forward 111

     5.1.2   Decode-and-forward 112

     5.1.3   Orthogonalize-and-forward 113

5.2   Increasing the Spectral Efficiency with Two-way Relaying 116

     5.2.1   Amplify-and-forward 117

     5.2.2   Decode-and-forward 118

     5.2.3   Orthogonalize-and-forward 120

5.3   Increasing the Spectral Efficiency with Two-path Relaying 123

     5.3.1   Two-path Relaying for Fast Fading Channels 123

         5.3.1.1   Amplify-and-forward 124

         5.3.1.2   Decode-and-forward 129

     5.3.2   Two-path Relaying for Slow Fading Channels 131

5.4   Simulation Results 136

     5.4.1   Two-way Relaying 136

     5.4.2   Two-path Relaying 138

6   Spectral Efficient Signaling in Rank-deficient MIMO Channels by the Use of Relays   145

6.1   Introduction 145

6.2   Amplify-and-forward Protocols 148

     6.2.1   Signal and Channel Model 149

6.2.2   Achievable Rates 152

6.2.3   System with Distributed Relay Array (DRA) 154

6.2.4   System with Ad Hoc Relays 157

6.2.5   Numerical Examples 158

6.3   Ergodic Performance of a Relay-assisted Rank-deficient MIMO Channel 161

6.3.1   System Model 162

6.3.2   Achievable Rate 163

6.3.3   Numerical Results 166

7   Distributed MIMO Signaling in a Multi-hop Relay Channel   **171**

7.1   Introduction 171

7.1.1   Related Work 172

7.1.2   System Model 172

7.1.3   Signal and Channel Model 173

7.2   A Distributed MIMO System with Multi-hop Amplify-and-Forward Relays . . 173

7.3   A Distributed MIMO System with Multi-hop Decode-and-Forward Relays . . . 175

7.3.1   System Model 175

7.3.2   Achievable Rates 176

7.3.2.1   Weakest Link Signaling 177

7.3.2.2   Ergodic Signaling 180

7.4   Numerical Examples 181

8   Conclusions and Outlook   **185**

Acronyms   **195**

Notation   **199**

Bibliography   **201**

Curriculum Vitae   **214**

1 Introduction

1.1 Motivation

In recent years the interest in wireless ad hoc networks has grown due to their low deployment costs and the potential to build self-organizing heterogenous and pervasive wireless networks. The lack of infrastructure and the possibility of self-configuration lead to various commercial and military applications. Examples are wireless personal area networks [2], home networks [3] and wireless sensor networks [4]. An ad hoc wireless network is a collection of wireless mobile nodes that form a network without the aid of a prescribed infrastructure [5]. In contrast to cellular systems the mobiles handle the necessary networking tasks by themselves through the use of distributed protocols and control algorithms. Multi-hop connections, where intermediate nodes relay the message to the final destination are of mandatory importance to achieve connectivity, enhance transport capacity and power efficiency [6].

At the level of point-to-point wireless communication, a breakthrough was the idea to use multiple antennas at the transmitter and the receiver [7], [8], [9], [10] which made it possible to introduce spatial degrees of freedom into a wireless communication system. Space-time signal processing makes use of these degrees of freedom to boost capacity and/or to enhance the reliability of multiple-input multiple-output (MIMO) communication systems where a system with M transmit and N receive antennas constitutes a $M \times N$ MIMO channel. It was shown in [9] that for statistically independent Gaussian channel coefficients the ergodic capacity of a MIMO channel scales linearly with $\min\{M, N\}$ compared to a corresponding single-input single-output (SISO) channel [9]. With *spatial multiplexing* one can increase the data rate substantially without additional cost of bandwidth or power by transmitting data streams simultaneously over spatial sub-channels which are available in a rich scattering environment [10] or in a multi-mode fiber [11]. On the other side *space-time codes* are used to mitigate the channel fading effects by utilizing the spatial diversity of the MIMO channel [12]. In view of the cost of bandwidth these observations has initiated a large amount of research on MIMO systems in recent years. However, there is a major obstacle in the practical exploitation of these MIMO technologies: the capacity and diversity gains depend strongly on the propagation environment and diminishes with increasing correlation of the channel coefficients [13].

Cooperation between terminals at the physical layer (PHY) is the natural extension of space-time signal processing to multiple distributed nodes in an ad hoc network. The distributed nodes essentially form a virtual antenna array. Completely new possibilities arise due to the large number of nodes in dense networks that possibly could cooperate, the availability of signal processing capability at other nodes and the nature of the wireless medium, where information (signaling bits and data bits) between adjacent nodes can be exchanged without additional communication infrastructure. The most basic form of PHY node cooperation is linear *amplify-and-forward relaying* (AF). The relay receives in the first time slot the signal from the source and forwards an amplified version to the destination or to the next relay in the second time slot. This way of relaying leads to low-complexity relay transceivers and to lower power consumption since there is no signal processing for decoding procedures. Other forms of node cooperation are *decode-and-forward relaying* (DF) or *compress-and-forward relaying* (CF). In the DF strategy the relay decodes the data of the source and transmits a re-encoded version of the data to the destination. In CF relaying the relay quantizes its received signal from the source in an efficient way. It then transmits a codeword that represents the quantized observation and allows the destination to recover the quantized relay observation. These strategies will be discussed in full detail in the course of this thesis.

We are interested how *physical layer cooperation* between terminals in a wireless network can help to achieve higher spectral efficiencies, similar to MIMO systems. By cooperation we mean that distant terminals share their antenna resources, i.e., terminals help other terminals in the communication process. As mentioned, one form of cooperation is relaying, where a third terminal relays data from one terminal (*source*) to another terminal (*destination*). The capacity of this simple three-node relay channel is still not known, but tight upper and lower bounds are available [14], [15], [16], [17], [18], [19]. The first results about relay channels have been presented in the seventies [20], [21], [14], where the paper by Cover and Gamal [14] presents by far the most important work about relay channels. We will review the results of this paper in detail in Chapter 3. We will see that different relaying strategies lead to different gains that are comparable to gains achievable in classical MIMO systems. For example, the decode-and-forward scheme provides a transmit array gain and/or diversity gain (source and relay transmit the same symbol). The compress-and-forward strategy provides a receive array gain and/or diversity gain (relay and destination observe two independent observations of the same source symbol). We will see that practical relaying systems often assume that the relays can operate only in a half-duplex mode, i.e., can only receive *or* transmit in a given time and frequency slot, but not both simultaneously. However, this practical restriction leads to a severe loss in spectral efficiency as we will see later in this work and it seems, that the most prominent MIMO gain, the *multiplexing gain* cannot be achieved by half-duplex relay systems. In Chapter 5 we will propose relaying protocols that allow to use half-duplex relays and are able to achieve a high spectral efficiency.

The 80ties and 90ties didn't see much work in relay channels. However, the pioneering work done in [6], [22], [23], [24] [25], [26], [27] triggered the re-interest in relay channels and multi-hop channels and have lead to an enormous amount of papers in that area in the recent five, six years. The next section will discuss some of the papers that are relevant for this thesis.

1.2 General Overview of Related Work

As mentioned previously, the exploitation of spatial diversity to mitigate the effects of fading and therefore to increase the reliability of radio links in wireless systems is a well known technique for transceivers with co-located antennas (space-time coding) [12,28–31]. At the end of the 90ties new forms of spatial diversity has been introduced called *cooperative diversity* [24, 26, 27] or *user cooperation diversity* [22, 23]. The main idea is that terminals share their antenna resources to build a virtual macro antenna array, realizing spatial diversity in a distributed fashion. In such a network terminals can serve as relays for other terminals. Relaying has therefore become a major topic in the wireless research community during the last five years and it is considered to be an important and growing area of research for the years to come.

In short, a relay can help to increase the maximum data rate (capacity) and/or the robustness (diversity) of the wireless connection between a source terminal and destination terminal. Usually, the relays are classified according to the way they forward the signals from the source to the destination. The most important strategies are:

- amplify-and-forward (AF),
- decode-and-forward (DF),
- compress-and-forward (CF).

In the literature one can also find modifications and combinations of these strategies, e.g., *partial decode-and-forward* [14], [15], *decode-amplify-forward* [32], *bursty amplify-and-forward* [33] and so on. Also different names for the same strategies do often appear. For example *compress-and-forward* is often called *quantize-and-forward* or *estimate-and-forward*. Decode-and-forward is sometimes called *regenerative* relaying and amplify-and-forward is sometimes known as *nonregenerative* relaying.

In the AF strategy the relay amplifies its received signal according to its transmit power constraint (average or peak power constraint) and re-transmits a noisy version of the source signal to the destination. Note that the relay has not to decode anything and this leads to low-complexity relay transceivers and lower power consumption since there is no need of signal processing for decoding procedures. However, the noise at the relay is also forwarded to the destination. The deployment of many AF relays can lead to a large increase of the noise level at the destination, thereby making the AF strategy not useful in certain cases. An advantage of AF relays is that they need not to know the modulation and coding method used at the source and destination, i.e., AF relaying is suitable when adaptive modulation techniques are employed at the source. Note that there are also researchers, who consider AF to be an expensive strategy. The reason is, that building a completely analog AF relay may be a challenging task, i.e., a circuit which has an input $r(t)$ and produces an output $10 \cdot r(t)$. However, in practice it is often suitable to realize an AF relay semi-digitally, i.e., an analog-to-digital (ADC) converter transforms the received signal $r(t)$ at the relay into the digital domain where the amplification is done digitally. Afterwards the digital result is transformed back into the analog domain with a digital-to-analog (DAC) converter. For completely analog AF relays it has been proven in [27] that second order diversity can be achieved (with one relay). Strictly speaking, this result may not be applicable directly to a semi-digital AF relay, since there might occur read/write errors and quantization errors which are not present in an analog AF relay.

In contrast, a DF relay decodes the source data, re-encodes (using the same code book or a different code book) and forwards a fresh codeword to the destination. This strategy does not suffer form noise amplification (as the AF strategy) but from error propagation, i.e., decoding errors can occur at the destination as well as at the relay.

The CF strategy can be seen as a compromise between AF and DF. The relay quantizes its observation, uses a specific source code to compress the bits obtained from the quantization and sends to the destination the codeword that represents the quantized observation. We will discuss the three main strategies AF, DF, and CF in detail in Chapter 3. Literature on cooperative relaying can be divided into two classes:

i) the relay terminals act only as relays and do not have own data to transmit [14], [24, 26, 27],

ii) the relay terminals are user terminals as well and have own data to transmit [22, 23].

For example, in the latter case two users employ Markov based superposition coding to transmit own data and data from the other terminal to a base station. Such cooperation leads not only to an increase in the uplink capacity for both users but also to a more robust system, where the rates of the users are less sensitive to channel variations. This thesis deals only with schemes for the first class of cooperative relaying protocols. We therefore review only the most important literature in this area.

In [24,26,27] protocols for cooperative diversity to combat fading and shadowing are presented and analyzed in terms of outage probability. The authors consider half-duplex terminals, i.e., the devices are not able to transmit and receive at the same time and frequency. In [27], a cooperative transmission scheme with two transmit terminals T_1 and T_2 is considered. Both terminals have information to transmit to the same destination or to two different destinations. In the first time slot, T_1 broadcasts its symbol to its destination and T_2. In the second time slot, T_2 acts as relay and transmits the received symbol to the destination while T_1 keeps silent. Upon receiving the signal, the destination decodes the two versions of received signals using temporal maximum ratio combining. After T_1 has completed his first transmission, T_2 transmits in the same way and T_1 acts as relay. In the paper, fixed relaying schemes which comprise AF and DF are considered. In the AF scheme the relay simply amplifies what it receives with respect to a transmit power constraint. In the DF scheme the relay fully decodes the received signal, re-encodes it and forwards a fresh codeword to the destination. Selection relaying and incremental relaying schemes which can use either AF or DF relays are also proposed. In selection relaying, the source repeats its message in the second time slot if the SNR of the source to relay channel is below a certain threshold. It is an adaptive version of AF or DF that reduces to direct transmission with repetition coding if the relay cannot decode. In incremental relaying, the destination sends some feedback to the source and relay upon receiving the symbol from the source in the first time slot. If the destination can correctly decode the received symbol in the first time slot, the source can directly transmit the next message and no relaying is required. The mutual information of the presented schemes is analyzed in terms of outage probability $P\{I < R\}$, where I denotes the mutual information between source and destination and R is the fixed communication rate. An outage occurs if the mutual information I of the cooperative relaying link is below a certain target rate R. Using high SNR approximations, the authors calculate the diversity orders for each relay strategy. It is concluded that all cooperative diversity protocols, except the fixed DF scheme, achieve full diversity (second-order diversity in the case of one relay terminal). The performance of the fixed DF scheme, however, is limited by its source/relay path.

Similar scenarios have been presented in [34–40]. In [36–38] the mutual information and the

15

outage probability behavior of three different TDMA (time division multiple access) relaying protocols is analyzed. In protocol I the source transmits new information to relay and destination in the first time slot. In the second time slot the relay forwards the data to the destination whereas the source transmits new information (which is only received by the destination as superposition of the source and relay signal). This protocol was also proposed in [41, 42]. Protocol II is the same as in [27], which is equal to protocol I except that the source is quiet in the second time slot. In protocol III the source transmits information to the relay in the first time slot. Source and relay then form a virtual antenna in the second time slot and transmit the same information to the destination. It is shown that the mutual information for the three protocols in the case of AF relaying can be ordered as $I_1^{\mathrm{AF}} \geq I_2^{\mathrm{AF}} \geq I_3^{\mathrm{AF}}$, whereas in the case of DF relaying it can be ordered as $I_1^{\mathrm{DF}} \geq I_3^{\mathrm{DF}} \geq I_2^{\mathrm{DF}}$.

Work on coded cooperation between two users (user cooperation diversity) is presented, e.g., in [43–47] by Hunter et al., in [48–50] by Stefanov et al. and in [51, 52] by Agustin et al. These schemes belong to the second class of cooperative diversity schemes as mentioned above. In [47], e.g., an uplink communication with two users is considered. Each user has its own data to transmit. The users encode blocks of K source bits using a cyclic redundancy check (CRC) code concatenated with a forward error correcting (FEC) code, so that, for an overall rate R code, we have $N = K/R$ total code bits per source block. Two users cooperate by dividing the transmission of their N-bit codewords into two successive time segments, or frames. In the first frame, each user transmits a rate R_1 code word with $N_1 = K/R_1$ bits, where $R_1 > R$. This can be viewed as a subset of the total N allocated code bits that contains all the original information. Each user also receives and decodes the partners transmission. If the user successfully decodes the partners rate R_1 codeword (by checking the CRC bits), the user computes and transmits N_2 additional parity bits for the partners data in the second frame, where $N_1 + N_2 = N$. These additional parity bits are selected such that they can be combined with the codeword in the first frame to produce a more powerful rate R codeword. If the user does not successfully decode the partner, N_2 additional parity bits for the users own data are transmitted. Each user always transmits a total of N bits per source block over the two frames, and the authors define the level of cooperation as $\alpha = N_1/N$. Coded cooperation can use a wide variety of channel coding methods. For example, the overall code may be a block or convolutional code, or a combination of both. The code symbols for the two frames may be partitioned through puncturing, product codes, or other forms of concatenation. A good overview about coded cooperation schemes can be found in [53].

AF and DF relaying schemes which do not consider a direct link between source and destination are presented in [54–61]. These schemes can be classified as pure relaying schemes since only one signal contribution arrives at the destination and therefore no "cooperation" in form of coherent superposition[1] at the destination is possible. The end-to-end performance of a two-hop communication scheme subject to Rayleigh fading is presented in [55,56]. For AF and DF the probability density function (PDF) and the cumulative distribution function (CDF) of the SNR at the destination are derived. Based on this the outage probability and average BER is analyzed. It is shown that DF performs better than AF in terms of outage probability in the low SNR regime, whereas both schemes converge in the high SNR regime. This analysis is extended to Nakagami fading channels for AF relaying in [57]. In [58] the optimal power allocation among source and relay is provided to minimize the outage probability for AF and DF relaying. These papers have considered an amplification gain at the relay which ensures a certain instantaneous transmit power at the AF relay given by

$$g = \sqrt{\frac{P_r}{P_s \alpha_1^2 + \sigma_r^2}}, \tag{1.1}$$

where α_1 is the fading amplitude of the channel between source and relay, P_s is the source transmit power, P_r the relay transmit power and σ_r^2 the relay noise variance. In [59] the performance of a "fixed gain relay" is analyzed. A "fixed gain relay" only ensures an average transmit power, i.e.,

$$g = \sqrt{\frac{P_r}{P_s \exp\left(\alpha_1^2\right) + \sigma_r^2}}. \tag{1.2}$$

Numerical results show that these low complexity relays have comparable performance to AF relaying systems with variable gain. Further it is shown that relay amplifier saturation causes only a minimal loss in performance of these systems.

Up to now we discussed only the case where one single antenna relay facilitated the communication between one source and one destination. Therefore, the diversity order which is achievable is limited to two. One way to increase the diversity order is to use multiple antennas at source and relay. Another way is to use multiple single antenna relays. In this case the relays form either a multiple antenna array in the second slot utilizing a space-time code in the same channel or retransmit the signals on orthogonal channels to the destination. The latter case causes a bandwidth efficiency reduction of factor $1/K$ if K relays participate in the cooperation. Therefore, using a space-time code in the same channel is favorable in terms of bandwidth efficiency. In [62, 63] Laneman et al. extended their results from [24, 26, 27] to the case of

---

[1]Achieving coherent superposition of the source signal and the relay signal at the destination is one form of cooperation and is explained in more detail in Chapter 3.

multiple relays. The authors consider a wireless network with m transmit terminals that form a set $\mathcal{M}$. Each source s in $\mathcal{M}$ has information to transmit to its destination $d_s \notin \mathcal{M}$, potentially using the other terminals as relays. The authors considered a repetition-based cooperative diversity algorithm and a space-time-coded cooperative diversity algorithm to transmit the source information to the destination. In the first phase, the source broadcasts to its destination and all potential relays. During the second phase, the other terminals relay to the destination, either on orthogonal subchannels in the case of repetition-based cooperative diversity, or simultaneously on the same subchannel in the case of space-time-coded cooperative diversity.

A distributed implementation of the Alamouti space-time coding scheme using AF and DF relays is presented in [64–66]. In the distributed AF case this scheme is not able to make the effective channel orthogonal, but it still achieves a degree of diversity which is between two and three (the system consists of one source-destination pair and two relays). The DF case suffers form error propagation. It is shown that DF is always slightly better than AF in terms of average symbol error probability. Unfortunately, full rate orthogonal space-time block codes for more than two antennas which can be assigned to cooperative relay networks (more than two relays) are not available. The performance of higher order orthogonal space-time block codes is presented in [67, 68].

A space-time coding scheme for multiple AF relays was presented in [69]. The main idea of the scheme is to transform the spatial diversity which is inherent in such a distributed network into the temporal domain. This is done by time-variant and relay specific signal processing at the relays. This signal processing can be either a multiplication of the received signal with a *phase signature sequence* or *relay switching*. In the case of phase multiplication the phase signature sequences for each relay are favorably derived from the columns of an FFT matrix. In the case of relay switching the relays are switched one after the other on and off such that only one relay is active at one time instance. As consequence of this time-variant signal processing the effective channel seen by the destination is time-variant although the physical channel impulse responses are time-invariant. The temporal diversity can then be exploited by a linear block code employed at the source.

In [70, 71] the behavior of diversity for space-time coded cooperative diversity using multiple fixed gain AF relays is analyzed. The relay nodes encode their received signals into a *distributed linear dispersion code* [72], and then transmit the coded signals to the receive node. The main result is that the diversity of the system behaves as

$$\min(T, K) \left(1 - \frac{\log \log P}{\log P} \right),$$

with T the coherence interval of the channel, K the number of relay nodes, and P the total transmit power. It shows that when $T \geq K$ and the average total transmit power is very high ($P \gg \log P$), the relay network has almost the same diversity as a multiple-antenna system with K transmit antennas, which is the same as assuming that the K relay nodes can fully cooperate and have full knowledge of the transmitted signal. In other words, for high SNR the pairwise error probability (PEP) behaves as $\left(\frac{\log P}{P}\right)^{\min(T,K)}$. Thus, apart from the $\log P$-factor and assuming $T \geq K$, the system has the same diversity as a multiple-antenna system with K transmit antennas. It is further shown that, assuming $K = T$, the leading order term in the PEP behaves as

$$\frac{1}{|\det(\mathbf{S}_i - \mathbf{S}_j)|^2} \left(\frac{8 \log P}{P}\right)^K,$$

which compared to the PEP of a space-time code:

$$\frac{1}{|\det(\mathbf{S}_i - \mathbf{S}_j)|^2} \left(\frac{4}{P}\right)^K,$$

shows the loss of performance due to the fact that the code is implemented distributively and the relay nodes have no knowledge of the transmitted symbols. At low and high SNR, the coding gain is the same as that of a multiple-antenna system with K antennas. However, at intermediate SNR, it can be quite different, which has implications for the design of distributed space-time codes.

So far we gave a broad overview about diversity results in relay channel. We now turn our attention to capacity results and summarize the main contributions. A more detailed review of the most important capacity results will be given in Chapter 3. First results about the capacity of the relay channel haven been found in the seventies by [20], [73] and [14]. The capacity problem for the general relay channel is still unsolved. However, for the following special cases the capacity is known [14], [74]:

- Degraded relay channel,

- Reversely degraded relay channel,

- Relay channel with feedback,

- Deterministic relay channel.

In [14] upper and lower bounds for the general relay channel are presented. For the upper bound the authors used a max-flow min-cut argument [74, Theorem 14.10.1]. For the lower bounds the authors propose two relaying strategies: *decode-and-forward* and *compress-and-forward*. We will discuss the results and the techniques in more detail in Chapter 3. In [15, 16, 75–77] the capacity results have been extended to the fading relay channel. The main result is that for certain geometric constellations, the capacity of the general fading relay channel is known. In particular, when the relay is near to the source the upper bound coincides with the decode-and-forward lower bound. When the relay is near to the destination, the upper bound coincides with the compress-and-forward lower bound. More details are given in Chapter 3.

Despite the recent developments, the capacity of the general (wireless) relay channel is still not found. Concerning that, the paper [6] by Gupta and Kumar lead to a paradigm shift. Instead of trying to find the exact capacity (region) of networks with finite number of terminals (as the classical three-node relay channel), the authors proposed to let the number of terminals go to infinity and to study the scaling behavior of the capacity [78], [79]. The main result of the paper is that the transport capacity[2] scales as $\Theta(\sqrt{N})$ as the number of terminals N tends to infinity. This means that the the per-node throughput scales as $\Theta(1/\sqrt{N})$. The results were obtained by consideration of sub-optimal receivers, i.e., the interference in the network was not exploited information-theoretically. The follow-up paper [80] analyzed the scaling behavior from an information-theoretic viewpoint with the result that the transport capacity scales as $\Theta(N)$ for large N, i.e., a per-node capacity of $\Theta(1)$ is possible. It was shown then in [81, 82] that the capacity of a network with one source-destination pair and multiple relays scales as $\log N$ when N is the number of terminals. This result says that the capacity of the multiple relay channel[3] is known asymptotically.

Further recent work about the capacity gains of relay channels can be found in [33, 77, 83–89] and the reference therein. The paper [90] studies bounds on the capacity of MIMO relay channels.

1.3 Main Contributions of the Thesis

After studying the techniques developed for classical relay channels we extend theses protocols to the *two-way relay channel* and provide achievable rate regions. We show that in an AWGN relay channel a combined strategy of compress/decode-and-forward achieves the cut-set upper bound on the sum-capacity of the two-way relay channel, when the relay is near to one of the terminals. In a phase fading relay channel a pure compress-and-forward strategy achieves the

---

[2]The transport capacity quantifies the number of bits that are transported *one* meter towards its destination in *one* second.

[3]Note that no assumption of degradeness or feedback is necessary. The result is valid for general relay channels in the asymptotic limit of large number of relays.

cut-set upper bound when the relay terminal is near to one of the terminals. The results are presented in Chapter 4 and have been published in

1. Boris Rankov and Armin Wittneben, "**Achievable Rate Regions for the Two-way Relay Channel**", International Symposium on Information Theory (ISIT), Seattle, USA, July 2006.

In Chapter 5 we provide two solutions to the *factor one-half problem* in half-duplex relay channels [63], [60]. We propose *two-way relaying* and *two-path relaying* in order to increase the pre-log factor. In two-path relaying two relays alternate in transmission and reception. All nodes operate in the same physical channel, thereby causing inter-relay interference. For amplify-and-forward (AF) relays, it is shown that this protocol can recover a significant portion of the half-duplex loss (pre-log factor $\frac{1}{2}$) for weak to moderately strong inter-relay channels. For DF relays the half-duplex loss is negligible when the inter-relay channel is very weak or very strong and reasonable medium inter-relay channels (three different DF strategies are applied, depending on the strength of the inter-relay channel). In two-way relaying[4], we propose a relaying protocol where a synchronous bidirectional connection between two half-duplex terminals is established using one half-duplex relay. Hereby, the achievable rate in one direction suffers still from the pre-log factor $\frac{1}{2}$ but since two connections are realized in the same physical channel the sum-rate is considerably higher. We show that for DF relays the cut-set upper bound is achieved, when the relay is in the proximity of one of the terminals. This work has been published in

1. Boris Rankov and Armin Wittneben, "**Spectral Efficient Protocols for Nonregenerative Half-duplex Relaying**", Forty-third Annual Allerton Conference on Communication, Control, and Computing, Allerton (IL), USA, Oct. 2005,

2. Boris Rankov and Armin Wittneben, "**Spectral Efficient Signaling for Half-duplex Relay Channels**", Asilomar Conference on Signals, Systems, and Computers, Pacific Grove (CA), USA, Nov. 2005,

3. Boris Rankov and Armin Wittneben, "**Spectral Efficient Protocols for Half-duplex Fading Relay Channels**", IEEE Journal on Selected Areas in Communications, Feb. 2007, to appear.

We then study two applications of relays for MIMO communications. In Chapter 6 we propose a spectral efficient *relay-assisted signaling for rank-deficient MIMO channels*. Rank-deficient MIMO channels fail to provide significant multiplexing gains [9], [13]. We propose to use half-duplex AF relays as active scatterer in a way that the eigenvalue structure of the compound

---

[4]We apply the theory developed in Chapter 4 to the half-duplex case.

two-hop matrix channel improves with respect to the multiplexing capability. The resulting factor one-half loss in spectral efficiency due to the half-duplex operation of the relays can be circumvented by the protocols proposed in Chapter 5. The results have been published in

- Armin Wittneben and Boris Rankov, "**Impact of Cooperative Relays on the Capacity of Rank-Deficient MIMO Channels**", 12th IST Summit on Mobile and Wireless Communications, Aveiro, Portugal, June 2003,

- Boris Rankov and Armin Wittneben, "**On the Capacity of Relay-Assisted Wireless MIMO Channels**", Signal Processing Advances in Wireless Communications (SPAWC), Lissabon, Portugal, July 2004,

- Boris Rankov, Jörg Wagner and Armin Wittneben, "**Distributed Antenna Systems and Linear Relaying for Rank-Deficient MIMO Systems**", Chapter in Distributed Antenna Systems: Open Architecture for Future Wireless Communications, Auerbach Publications, CRC Press, 2007, to appear,

- Jörg Wagner, Boris Rankov and Armin Wittneben, "**Replica Analysis of Correlated MIMO Relay Channels**", IEEE Transactions on Information Theory, 2006, submitted,

and parts of the results have been published in [91], [92], [93].

The second application focuses on multi-hop relay systems. We propose two systems, that are capable to achieve a distributed *spatial* multiplexing gain with single-antenna nodes. The first system uses AF relays and it is shown that the achievable rate decreases with increasing number of hops due to two effects: accumulation of relay noise along the multi-hop chain and singularity of the product channel matrix. The second system uses DF relays, where it is shown that for a weakest link signaling protocol, the achievable rate also decreases with increasing number of hops, although there is no noise accumulation along the multi-hop chain. For an ergodic signaling protocol the achievable rate is independent of the number of hops. The results have been published

- Boris Rankov and Armin Wittneben, "**Distributed Spatial Multiplexing in a Wireless Network**", Asilomar Conference on Signals, Systems, and Computers, Pacific Grove (CA), USA, Nov. 2004.

1.4 Organization of the Thesis

Chapter 3 introduces the classical relay channel. We review the results that were obtained in the late seventies for general relay channels and AWGN channels and present recent results about fading relay channels. Chapter 4 presents the extension of the relaying strategies discussed in Chapter 3 to *two-way* relay communication. Chapter 5 presents two relaying strategies that circumvent the factor $\frac{1}{2}$ problem in half-duplex relaying. The following two chapters present two applications of relaying. In Chapter 6 we propose to use relays for MIMO communication in rank-deficient channels. Chapter 7 proposes a multi-hop network of single-antenna nodes in order to realize a distributed MIMO system. The appendix contains some proofs of Theorems provided in this work.

2 Information-theoretic Preliminaries

This chapter provides an introduction into information-theoretic concepts and tools that are useful for relay channels. We start by introducing the theory of typical sequences which is the main tool for proving rate achievability of different transmission strategies for relay channels (Chapter 3) and two-way relay channels (Chapter 4). Next we discuss the application of typical sequences to lossless source coding and explain the concept of *binning*. We then look at lossy source coding without (Rate Distortion Theory) and with (Wyner-Ziv coding) side information at the decoder. These concepts will be applied in the derivation of the compress-and-forward strategy discussed in Chapters 3 and 4.

We assume that the reader is familiar with the basic information-theoretic terms and concepts, such as *Entropy*, *Relative Entropy*, *Conditional Entropy*, *Mutual Information* and *Capacity* as defined and presented in [74].

2.1 Theory of Typical Sequences

Typical sequences were introduced by Shannon in his 1948 paper "A Mathematical Theory of Communication" to prove capacity theorems based on random coding arguments. A very good introduction to the theory of typical sequences is given in the ETH lecture notes of James Massey [94]. In this chapter we summarize the most important definitions and properties of typical sequences.

2.1.1 Weak vs. Strong Typicality

There exist several definitions of typical sequences. We note from [94]: "At least half a dozen precise definitions of typical sequences have been given since 1948 ... The choice of definition for typical sequences is more a matter of taste than of necessity". However, weak typicality as defined in [74, Chapter 3] applies to continuous as well as to discrete random variables, whereas strong typicality as defined in [74, p.288] applies only to discrete random variables.

Weak typicality requires that the empirical entropy of a sequence is close to the true entropy, whereas *strong typicality* requires that the empirical distribution of a sequence is close to the true distribution. The definition of strongly typical sequences is more intuitive and includes the definition of weakly typical sequences for discrete random variables. Moreover, the proof of the compress-and-forward (CF) strategy requires strong typicality to invoke the Markov lemma. The caveat is that strong typicality does not apply to continuous random variables, i.e., achievability results obtained for discrete alphabets may not be valid for continuous alphabets. However, for Gaussian distributed random variables (which will be assumed for the CF strategy) the Markov lemma can be generalized and the results remain valid [15, Remark 30], [95]. As common in the relaying literature we will therefore use strong typicality for derivations and we refer to [95] when we apply the results to the case of continuous AWGN channels. A comprehensive treatment of strongly typical sequences is provided in [96].

2.1.2 Definition and Properties of Strongly Typical Sequences

A sequence $x^n = (x_1, x_2, \ldots, x_n)$ where $X_i \in \mathcal{X}$ are i.i.d. with distribution $P_X(\cdot)$, is said to be *strongly typical* with respect to $P_X(\cdot)$ if

$$
\begin{aligned}
&\left| \tfrac{1}{n} N(a|x^n) - P_X(a) \right| < \tfrac{\epsilon}{|\mathcal{X}|} \quad &&\text{for all } a \in \mathcal{X} \text{ with } P_X(a) > 0 \\
&N(a|x^n) = 0 &&\text{if } P_X(a) = 0
\end{aligned}
\tag{2.1}
$$

for any $\epsilon > 0$, where $\mathcal{X}$ is the finite alphabet set and $N(a|x^n)$ denotes the number of occurrences[1] of the letter a in the sequence x^n. Strongly typical sequences are therefore sequences whose empirical probability distribution $1/n \cdot N(a|x^n)$ is close to the true distribution $P_X(a)$ for all $a \in \mathcal{X}$. The law of large numbers says that the empirical distribution converges to the true distribution when n grows to infinity, i.e., for a fixed ϵ condition (2.1) can be achieved by increasing n correspondingly. The division by the size of $\mathcal{X}$ in (2.1) is due to technical reasons and is useful for various bounds based on (2.1). The second condition in (2.1) ensures that no letters with zero probability occur in ϵ-typical sequences.

The set of sequences x^n satisfying (2.1) is called the *strongly typical set* $T_\epsilon^n(P_X)$ with respect to ϵ and $P_X(\cdot)$. The following theorem summarizes three important properties of strongly typical sequences [97].

---

[1]Note that the cardinality of the alphabet of a continuous random variable is infinite and uncountable and therefore this definition is not applicable.

Theorem 2.1.1. *Suppose that a sequence X^n is emitted[2] by a discrete memoryless source[3] (DMS) with letter probability $P_X(\cdot)$ and suppose that $x^n \in T_\epsilon^n(P_X)$. We have*

$$2^{-n(H(X)+\epsilon_1)} \le P_X^n(x^n) \le 2^{-n(H(X)-\epsilon_1)} \tag{2.2}$$

$$(1 - \epsilon_2(n))2^{n(H(X)-\epsilon_1)} \le |T_\epsilon^n(P_X)| \le 2^{n(H(X)+\epsilon_1)} \tag{2.3}$$

$$\Pr[X^n \in T_\epsilon^n(P_X)] \to 1 \text{ as } n \to \infty \tag{2.4}$$

where $\epsilon_1 = f(\epsilon) \to 0$ as $\epsilon \to 0$ and $\epsilon_2(n) \to 0$ for fixed ϵ and $n \to \infty$. We denote by $P_X^n(x^n) = \prod_{i=1}^n P_X(x_i)$ the probability that the sequence x^n is put out by the source $P_X(\cdot)$.

The second relation (2.3) says that the number of typical sequences $|T_\epsilon^n(P_X)|$ in the set of all possible sequences $\mathcal{X}^n$ is[4] approximately $2^{nH(X)}$. The first relation (2.2) says that all typical sequences are equiprobable and that the probability is approximately $2^{-nH(X)}$. Finally, the third relation (2.4) says that the probability observing a typical sequence of length n as output of a DMS is nearly one for large n.

Proof of Theorem 2.1.1. First note that the first part of (2.1) can be rewritten as

$$n\left(P_X(a) - \frac{\epsilon}{|\mathcal{X}|}\right) \le N(a|x^n) \le n\left(P_X(a) + \frac{\epsilon}{|\mathcal{X}|}\right) \tag{2.5}$$

for all $a \in \text{supp}(P_X)$ where $\text{supp}(f)$ denotes the support of the function $f(x)$, i.e., the set of all x where $f(x) \ne 0$. We prove the upper bound in (2.2). For $x^n \in T_\epsilon^n(P_X)$ we have

$$P_X^n(x^n) = \prod_{a \in \text{supp}(P_X)} P_X(a)^{N(a|x^n)} \tag{2.6}$$

$$\le \prod_{a \in \text{supp}(P_X)} P_X(a)^{n(P_X(a) - \epsilon/|\mathcal{X}|)} \tag{2.7}$$

$$= 2^{\sum_{a \in \text{supp}(P_X)} n(P_X(a) - \epsilon/|\mathcal{X}|) \log P_X(a)} \tag{2.8}$$

$$= 2^{n\left(\sum_{a \in \text{supp}(P_X)} P_X(a) \log P_X(a) - \frac{\epsilon}{|\mathcal{X}|} \log P_X(a)\right)} \tag{2.9}$$

$$= 2^{-n\left(H(X) + \frac{\epsilon}{|\mathcal{X}|} \sum_{a \in \text{supp}(P_X)} \log P_X(a)\right)} \tag{2.10}$$

$$\le 2^{-n\left(H(X) + \frac{\epsilon}{|\mathcal{X}|}|\mathcal{X}| \log P_{\min}\right)} \tag{2.11}$$

$$= 2^{-n(H(X) - \epsilon_1)} \tag{2.12}$$

---

[2]We denote by X^n a length n sequence of random variables and by x^n a particular realization of X^n.

[3]Which means that the X_i in $X^n = (X_1, X_2, \ldots, X_n)$ are i.i.d. and that the alphabet $\mathcal{X}$ is discrete (but not necessarily finite).

[4]The set $\mathcal{X}^n$ contains all possible length n sequences x^n (typical and nontypical). For example, if the alphabet $\mathcal{X} = \{0, 1\}$ is binary, the cardinality of $\mathcal{X}^n$ is 2^n.

where $\epsilon_1 = -\epsilon \log(P_{\min})$ with $P_{\min}$ the smallest nonzero probability of $P_X(\cdot)$. Inequality (2.7) follows from the lower bound in (2.5) and the fact that $0 < P_X(a) \leq 1$ for all a. The derivation of the the lower bound in (2.2) follows similarly:

$$P_X^n(x^n) = \prod_{a \in \text{supp}(P_X)} P_X(a)^{N(a|x^n)} \tag{2.13}$$

$$\geq \prod_{a \in \text{supp}(P_X)} P_X(a)^{n(P_X(a)+\epsilon/|\mathcal{X}|)} \tag{2.14}$$

$$\geq 2^{-n(H(X)+\epsilon_1)}. \tag{2.15}$$

We prove now the bounds (2.3) on the size of the typical set $T_\epsilon^n(P_X)$. We write

$$1 = \sum_{x^n \in \mathcal{X}^n} P_X^n(x^n) \tag{2.16}$$

$$= \sum_{x^n \in T_\epsilon^n(P_X)} P_X^n(x^n) + \sum_{x^n \notin T_\epsilon^n(P_X)} P_X^n(x^n) \tag{2.17}$$

$$\geq \sum_{x^n \in T_\epsilon^n(P_X)} P_X^n(x^n) \tag{2.18}$$

$$\geq |T_\epsilon^n(P_x)| 2^{-n(H(X)+\epsilon_1)} \tag{2.19}$$

where the second inequality follows by using (2.15). The upper bound on the size of the typical set is then

$$|T_\epsilon^n(P_x)| \leq 2^{n(H(X)+\epsilon_1)}. \tag{2.20}$$

For the lower bound in (2.3) it is convenient to prove first the third property (2.4) of typical sequences. We write

$$\Pr[X^n \notin T_\epsilon^n(P_X)] = \Pr\left[\bigcup_{a \in \mathcal{X}} \left\{\left|\frac{N(a|X^n)}{n} - P_X(a)\right| \geq \frac{\epsilon}{|\mathcal{X}|}\right\}\right] \tag{2.21}$$

$$\leq \sum_{a \in \mathcal{X}} \Pr\left[\left|\frac{N(a|X^n)}{n} - P_X(a)\right| \geq \frac{\epsilon}{|\mathcal{X}|}\right] \tag{2.22}$$

$$\leq 2|\mathcal{X}| \cdot 2^{\frac{-2n\epsilon^2}{|\mathcal{X}|^2 \ln 2}} \tag{2.23}$$

$$:= \epsilon_2(n) \tag{2.24}$$

where the first inequality follows by the union bound and the second inequality by applying first the Chernoff bound and then the Pinsker inequality, see [97] for a detailed derivation. For

$n \to \infty$ we have $\epsilon_2(n) \to 0$ and therefore (2.4). We proceed with the proof of the lower bound in (2.3). We write

$$1 - \epsilon_2(n) \leq \Pr\left[X^n \in T_\epsilon^n(P_X)\right] \tag{2.25}$$

$$\leq \sum_{x^n \in T_\epsilon^n(P_X)} 2^{-n(H(X)-\epsilon_1)} \tag{2.26}$$

$$= |T_\epsilon^n(P_X)| 2^{-n(H(X)-\epsilon_1)} \tag{2.27}$$

and hence

$$|T_\epsilon^n(P_X)| \geq (1 - \epsilon_2(n)) 2^{n(H(X)-\epsilon_1)} \tag{2.28}$$

which concludes the proof of Theorem 2.1.1.

2.1.3 Jointly and Conditionally Typical Sequences

Theorem 2.1.1 can be generalized to include *joint typicality* and *conditional typicality*. Two sequences $x^n = (x_1, x_2, \ldots, x_n)$ and $y^n = (y_1, y_2, \ldots, y_n)$ with $x_i \in \mathcal{X}$ and $y_i \in \mathcal{Y}$ and where (X_i, Y_i) are i.i.d. with distribution $P_{XY}(\cdot, \cdot)$ are called *jointly typical* with respect to $P_{XY}(\cdot, \cdot)$ if

$$\begin{aligned}
\left|\tfrac{1}{n}N(a,b|x^n,y^n) - P_{XY}(a,b)\right| &< \tfrac{\epsilon}{|\mathcal{X}||\mathcal{Y}|} \quad \forall (a,b) \in \mathcal{X} \times \mathcal{Y} \text{ with } P_{XY}(a,b) > 0 \\
N(a,b|x^n,y^n) &= 0 \qquad\qquad\quad \text{if } P_{XY}(a,b) = 0.
\end{aligned} \tag{2.29}$$

Joint typicality of x^n and y^n, i.e., $(x^n, y^n) \in T_\epsilon^n(P_{XY})$ implies also marginal typicality, i.e., $x^n \in T_\epsilon^n(P_X)$ and $y^n \in T_\epsilon^n(P_Y)$. We show here that $x^n \in T_\epsilon^n(P_X)$, the fact that $y^n \in T_\epsilon^n(P_Y)$ is proved in the same way:

If $(x^n, y^n) \in T_\epsilon^n(P_{XY})$, then according to the definition of joint typicality given in (2.29) we have

$$\sum_a \sum_b \left|\frac{1}{n}N(a,b|x^n,y^n) - P_{XY}(a,b)\right| < \epsilon. \tag{2.30}$$

With

$$N(a|x^n) = \sum_b N(a,b|x^n,y^n) \tag{2.31}$$

we can write

$$\sum_a \left| \frac{1}{n} N(a|x^n) - P_X(a) \right| = \sum_a \left| \sum_b \frac{1}{n} N(a,b|x^n,y^n) - \sum_b P_{XY}(a,b) \right| \quad (2.32)$$

$$= \sum_a \left| \sum_b \left(\frac{1}{n} N(a,b|x^n,y^n) - P_{XY}(a,b) \right) \right| \quad (2.33)$$

$$\leq \sum_a \sum_b \left| \frac{1}{n} N(a,b|x^n,y^n) - P_{XY}(a,b) \right| \quad (2.34)$$

$$\leq \epsilon \quad (2.35)$$

where we used the triangle inequality in (2.34). Therefore, according to (2.1) we have $x^n \in T_\epsilon^n(P_X)$.

For a bivariate source that emits sequences (X^n, Y^n) the same relationships hold as in Theorem 2.1.1 when we replace $H(X)$ by $H(X,Y)$ and P_X by P_{XY}. It follows that there are approximately $2^{nH(X,Y)}$ typical (x^n, y^n) sequence pairs and about $2^{nH(X)}$ typical x^n sequences. Therefore for a given typical x^n, the number of y^n such that (x^n, y^n) is jointly typical is approximately

$$\frac{2^{nH(X,Y)}}{2^{nH(X)}} = 2^{nH(Y|X)} \quad (2.36)$$

on the average. The next theorem states that this is not only true in average, but is true for every typical x^n. For this we define

$$T_\epsilon^n(P_{XY}|x^n) = \{y^n : (x^n, y^n) \in T_\epsilon^n(P_{XY})\} \quad (2.37)$$

to be the set of all typical sequences $y^n \in T_\epsilon^n(P_Y)$ that are jointly typical with a given typical sequence $x^n \in T_\epsilon^n(P_X)$. The following theorem then generalizes Theorem 2.1.1 by including conditioning [97].

Theorem 2.1.2. *Suppose (X^n, Y^n) is emitted by a DMS $P_{XY}(\cdot, \cdot)$ and that $(x^n, y^n) \in T_\epsilon^n(P_{XY})$. We have*

$$2^{-n(H(Y|X)+\epsilon_1')} \leq P_{Y|X}^n(y^n|x^n) \leq 2^{-n(H(Y|X)-\epsilon_1')} \quad (2.38)$$

$$(1 - \epsilon_2'(n)) 2^{n(H(Y|X)-\epsilon_1')} \leq |T_\epsilon^n(P_{XY}|x^n)| \leq 2^{n(H(Y|X)+\epsilon_1')} \quad (2.39)$$

$$\Pr\left[Y^n \in T_\epsilon^n(P_{XY}|x^n) \,|\, X^n = x^n\right] \to 1 \ as \ n \to \infty \quad (2.40)$$

where $\epsilon_1' = f(\epsilon) \to 0$ as $\epsilon \to 0$ and $\epsilon_2'(n) \to 0$ for fixed ϵ and $n \to \infty$.

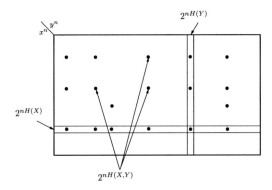

Fig. 2.1: Jointly and conditionally typical sequences

The proof of Theorem 2.1.2 is similar to the proof of Theorem 2.1.1 and is omitted here. See also [74, Excercise 13.10] for an outline of the proof. The properties of jointly and conditionally typical sequences can be nicely summarized in the two-dimensional array depicted in Fig.2.1. The vertical dimension represents all sequences x^n, and the horizontal dimension contains all y^n sequences. The rows represent all typical x^n sequences and there are approximately $2^{nH(X)}$ such rows (out of $|\mathcal{X}^n|$). The columns represent all typical y^n sequences and there are approximately $2^{nH(Y)}$ such columns (out of $|\mathcal{Y}^n|$). All sequence pairs (x^n, y^n) that are jointly typical are marked by a dot, there are approximately $2^{nH(X,Y)}$ dots (out of $|(\mathcal{X} \times \mathcal{Y})^n|$). In a given row (typical x^n sequence) there are about $2^{nH(Y|X)}$ dots, i.e., y^n sequences that are jointly typical with this specific x^n sequence. Similarly, there are $2^{nH(X|Y)}$ dots in each column.

The next theorem is at the heart of proving capacity theorems by random coding and typical sequence decoders.

Theorem 2.1.3. *Suppose $Y^n \in T_\epsilon^n (P_Y)$ is emitted by a DMS $P_Y(\cdot)$ and assume a fixed $x^n \in T_\epsilon^n (P_x)$. We have*

$$(1 - \epsilon_2'(n))2^{-n(I(X;Y)+\delta)} \leq \Pr\left[(x^n, Y^n) \in T_\epsilon^n (P_{XY})\right] \leq 2^{-n(I(X;Y)-\delta)} \qquad (2.41)$$

where $\delta = f(\epsilon) \to 0$ as $\epsilon \to 0$.

Before we discuss the intuition and implications of Theorem 2.1.3 we first prove it.

Proof of Theorem 2.1.3. For the upper bound we write

$$\Pr\left[(x^n, Y^n) \in T_\epsilon^n\left(P_{XY}\right)\right] = \sum_{y^n \in T_\epsilon^n(P_{XY}|x^n)} P_Y^n(y^n) \tag{2.42}$$

$$\leq \sum_{y^n \in T_\epsilon^n(P_{XY}|x^n)} 2^{-n(H(Y)-\epsilon_1)} \tag{2.43}$$

$$\leq 2^{n(H(Y|X)+\epsilon_1')} 2^{-n(H(Y)-\epsilon_1)} \tag{2.44}$$

$$= 2^{-n(I(X;Y)-\delta)} \tag{2.45}$$

where $\delta = \epsilon_1' + \epsilon_1$. In the first inequality we used (2.2) and in the second inequality (2.39). The lower bound follows accordingly:

$$\Pr\left[(x^n, Y^n) \in T_\epsilon^n\left(P_{XY}\right)\right] = \sum_{y^n \in T_\epsilon^n(P_{XY}|x^n)} P_Y^n(y^n) \tag{2.46}$$

$$\geq \sum_{y^n \in T_\epsilon^n(P_{XY}|x^n)} 2^{-n(H(Y)+\epsilon_1)} \tag{2.47}$$

$$\geq (1 - \epsilon_2'(n)) 2^{n(H(Y|X)-\epsilon_1')} 2^{-n(H(Y)+\epsilon_1)} \tag{2.48}$$

$$= (1 - \epsilon_2'(n)) 2^{-n(I(X;Y)+\delta)}. \tag{2.49}$$

Discussion of Theorem 2.1.3. As mentioned Theorem 2.1.3 is important for proving achievability results in the context of random coding and decoding based on typical sequences. Assume that the sequence x^n in Theorem 2.1.3 is a length n codeword transmitted through a discrete memoryless channel with probability distribution $p(y|x)$[5]. When n is large, the probability that the random codeword chosen at the transmitter is strongly typical goes to one as shown in (2.4). When such a codeword is transmitted through the channel the received sequence y^n will with high probability belong to one of the $2^{nH(Y|X)}$ sequences that are jointly typical with the transmitted codeword x^n with respect to $p(x, y)$. This is illustrated graphically in Fig. 2.2. As long as the sets of typical y^n sequences belonging to specific (typical) x^n sequences do not overlap the decoder can recover a transmitted x^n by searching the codebook for the x^n that is jointly typical with the received sequence y^n. Since the number of typical sequences y^n with respect to $p(y)$ is approximately $2^{nH(Y)}$ the number of codewords should not exceed

$$\frac{2^{nH(Y)}}{2^{nH(Y|X)}} = 2^{nI(X;Y)}. \tag{2.50}$$

---

[5]Because the channel is memoryless it is sufficient to consider single letter probabilities, since $p(y^n|x^n) = \prod_{i=1}^n p(y_i|x_i)$.

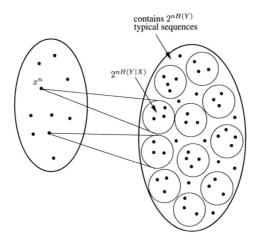

Fig. 2.2: Random coding and decoding based on typical sequences

Only then the sets containing the jointly typical sequences associated to each codeword do not overlap. This qualitative statement was made precise in Theorem 2.1.3.

For the compress-and-forward strategy that will be discussed in the following two chapters we need one additional Theorem and one Lemma. Conditional typicality can be generalized to [97]

Theorem 2.1.4. *Suppose that* $(u^n, y^n) \in T_\epsilon^n(P_{UY})$ *and that* $X_1, X_2, \ldots, X_n$ *are chosen independently using* $P_{X|U}(\cdot|u_i)$ *for* $i = 1, 2, \ldots, n$. *We have*

$$(1 - \epsilon'_2(n))2^{-n(I(X;Y|U)+\delta)} \leq \Pr\left[(u^n, y^n, X^n) \in T_\epsilon^n(P_{UYX})\right] \leq 2^{-n(I(X;Y|U)-\delta)} \quad (2.51)$$

where $\delta = f(\epsilon) \to 0$ *as* $\epsilon \to 0$.

The proof is along the same lines as the proof of Theorem 2.1.3. Referring to the compress-and-forward strategy for relay channels the sequence u^n corresponds to the transmitted codeword of the relay, the sequence x^n to the quantized observation of the relay and y^n corresponds to the received sequence at the relay. Further details will be provided in Chapters 3 and 4.

Furthermore, the proof of the compress-and-forward strategy requires a result concerning Markov chains X–Y–Z.[6]

---

[6] In the Markov chain given above the probability distribution of Z depends only on Y and is conditionally independent of X, see [74, Chapter 4] for a discussion of Markov chains.

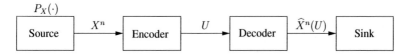

Fig. 2.3: The source coding problem

Lemma 2.1.5 (Markov Lemma). *Suppose that X–Y–Z is a Markov chain. For a fixed $(x^n, y^n) \in T_\epsilon^n (P_{XY})$, draw a z^n according to $p(z|y)$, which implies $\Pr\left[(y^n, Z^n) \in T_\epsilon^n (P_{YZ})\right] \rightarrow 1$ as $n \rightarrow \infty$. The Lemma states then that $\Pr\left[(x^n, y^n, Z^n) \in T_\epsilon^n (P_{XYZ})\right] \rightarrow 1$ as $n \rightarrow \infty$.*

For a proof of the Lemma we refer to [97] or [74, Lemma 14.8.1]. When we discuss the CF strategy we will see that X, Y and Z correspond to the destinations side information, the relay's receive signal and the relay's transmitted codeword, respectively. Note that for general random variables X, Y, Z when x^n and y^n are jointly typical and y^n and z^n are jointly typical we cannot conclude that x^n and z^n are jointly typical too, i.e., typicality is not transitive. However, the Markov lemma says, that typicality is transitive for Markov chains X–Y–Z, i.e., x^n and z^n are jointly typical too.

2.2 Lossless Source Coding

The properties of typical sequences can be exploited for efficient source coding (data compression). Consider the problem depicted in Fig.2.3. A discrete memoryless source emits a sequence x^n of length n and the source encoder maps the sequence into the index $u \in \{1, 2, \ldots, 2^{nR}\}$, where R is the rate of the compression code. The number of bits transmitted by the encoder may be constant or variable, depending on the source coding method. The decoder tries to recover the original source sequence $\widehat{x}^n(u) = x^n$ from the index u. We speak of lossless source coding or data compression when the probability of decoding error can be made zero by letting n growing to infinity. The rate R of the code is defined either as the average length of the codewords divided by the block length n for variable-length encoding or as k/n where k is the length of the codewords for fixed-length encoding.

We will look at two different possibilities to find short descriptions of n-length sequences. The first method (based on AEP[7]) indexes all source sequences belonging to the typical set and does not bother about the sequences outside the typical set. This method is an example

---

[7]AEP stands for *Asymptotic Equipartition Property* [74], which relates to the properties of typical sequences given in (2.2)–(2.4).

for variable-length source encoding. The second method (based on *Binning*) indexes all source sequences and rejects untypical sequences at the decoder side. This method uses fixed-length codewords.

2.2.1 Source Coding Based on AEP

Let $X^n = (X_1, X_2, \ldots, X_n)$ be a sequence of i.i.d. random variables drawn from the probability distribution $P_X(\cdot)$. We let $X_i \in \mathcal{X}$ where $\mathcal{X}$ is the discrete symbol set of finite cardinality. We would like to determine short descriptions for such sequences of random variables.

The set of all sequences is divided into two sets. The set $T_\epsilon^n(P_X)$ contains all typical sequences of length n and the complement set $\overline{T}_\epsilon^n(P_X)$ contains all nontypical sequences of length n. Each sequence in $T_\epsilon^n(P_X)$ is then represented by a unique index. Since the number of typical sequences in $T_\epsilon^n(P_X)$ is less or equal than $2^{n(H(X)+\epsilon)}$, cf. (2.3), the indexing requires no more than $n(H(X) + \epsilon) + 1$ bits[8]. All typical sequences are then prefixed by a 0, which leads to $n(H(X) + \epsilon) + 2$ bits describing these sequences (respectively, their indices). All nontypical sequences can be indexed by using no more than $n \log |\mathcal{X}| + 1$ bits. Each nontypical sequence is prefixed by a 1.

Instead of transmitting a specific sequence $x^n = (x_1, x_2, \ldots, x_n)$ directly, we transmit its index. The initial bit acts as flag bit and determines, whether the sequence is typical or not and therefore the length of the codeword. Note that codewords have two different lengths, depending whether the sequence to be transmitted is typical or not. Denoting by $l(x^n)$ the length of the codeword corresponding to x^n it can be shown [74, Chapter 3] that the expected length of the transmitted codewords satisfies

$$\mathcal{E}\left\{l(x^n)\right\} \leq n(H(X) + \epsilon'), \tag{2.52}$$

i.e., in average $nH(X)$ bits suffice (for large block lengths n) to describe X^n.

2.2.2 Source Coding Based on Binning

Consider again the sequence $X^n = (X_1, X_2, \ldots, X_n)$ of i.i.d. random variables drawn from the probability distribution $P_X(\cdot)$. For each sequence x^n choose an index at random from $\{1, 2, \ldots, 2^{nR}\}$, where $2^{nR} < |\mathcal{X}|^n$ (data compression[9]). All sequences X^n that belong to the

---

[8]The extra bit may be necessary in case $n(H(X) + \epsilon)$ is not an integer value.

[9]If $2^{nR} = |\mathcal{X}|^n$, then the number of indices is equal to the number of source sequences and therefore the number of bits needed to describe the indices is the same as the number of bits needed to describe the source sequences

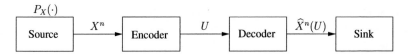

Fig. 2.4: Rate distortion and the source coding problem

same index form a *bin*. When the source emits a sequence x^n we transmit (or store) instead the corresponding bin index. The decoder which tries to recover the original x^n looks for a typical sequence within the bin. If there is one and only one typical sequence within the bin the decoder declares this sequence to be the estimate $\widehat{X}^n$ of the emitted source sequence, otherwise an error is declared. There are two types of error in this source coding approach: 1) if the source sequence is nontypical and 2) if the source sequence is typical but there is more than one typical sequence in the same bin. If the number of bins is much larger than the number of typical source sequences, the probability that more than one typical source sequence falls into the same bin is small. The goal is therefore to choose the number of bins (indices) so large that with high probability only one typical sequence falls into each bin, but that the number of bins is still smaller than the number of possible source sequences. It can be shown [74, Chapter 14, page 411] that the rate of the code R has to be larger than the entropy $H(X)$ in order to drive the error probability to zero as the block length goes to infinity, i.e.,

$$R > H(X) + \epsilon. \tag{2.53}$$

2.3 Lossy Source Coding or Rate Distortion Theory

We have seen that by exploiting the properties of typical sequences it is possible to find codes that allow to represent longer sequences by shorter sequences and still to recover the original, "longer" sequences with a probability of error going to zero as long as the rate of the compression code is larger than the entropy of the source that produces the sequences, i.e,

$$R > H(X), \tag{2.54}$$

where R is defined as the average (or fixed) length of the codewords divided by the block length n^{10}. We can ask the question whether we can reduce the rate of the code (i.e., increase the

---

$\rightarrow$ no data compression.

[10]Recall that block length n means that length n source sequences are mapped to codewords of fixed length or of variable length.

compression level) when we allow the decoder not to recover the original source sequence perfectly but within some specified distortion measure. The answer is positive and the theory behind it is known as *Rate Distortion Theory*.

Consider again the source coding problem depicted in Fig. 2.4. A discrete memoryless source (DMS) emits a sequence x^n. The encoder maps the sequence x^n to a codeword $\hat{x}^n(u)$ (quantization of x^n) where $u \in \{1, 2, \ldots, 2^{nR}\}$ and sends the index u to the decoder. The decoder puts out $\hat{x}^n(u)$ as the reconstruction (estimate) of x^n. The goal here is that the decoder reconstructs x^n only within some specified distortion measure, i.e., the reconstruction has not to be perfect (lossless) as in the previous section. By allowing for imperfect reconstruction one can hope to make the rate R of the code even smaller than $H(X)$. Rate distortion theory quantifies the tradeoff between the achievable rate R and distortion D.[11]

The choice of the distortion measure depends on the application and can be defined in several ways. For a DMS a natural choice is the *Hamming distance*, i.e.,

$$d^n(x^n, \hat{x}^n) = \frac{1}{n} \sum_{i=1}^{n} d(x_i, \hat{x}_i) \tag{2.55}$$

where $d(x, \hat{x}) = 0$ if $x = \hat{x}$, and $d(x, \hat{x}) = 1$ if $x \neq \hat{x}$. For sources with continuous alphabets one might define the *mean square error* as distortion measure

$$d^n(x^n, \hat{x}^n) = \frac{1}{n} \sum_{i=1}^{n} (x_i - \hat{x}_i)^2. \tag{2.56}$$

We present now a random code construction and show which rate R can be achieved for a given distortion measure D.

We first form a "quantization" codebook at the encoder as follows:

- Choose a distribution $P_{\hat{X}|X}$ freely

- Compute $P_{\hat{X}}$ as the marginal distribution of $P_{\hat{X}X}$[12]

- Generate 2^{nR} codewords $\hat{x}^n(u)$, $u \in \{1, 2, \ldots, 2^{nR}\}$, by choosing each of the $n \cdot 2^{nR}$ symbols i.i.d. according to $P_{\hat{X}}$.

Suppose now that a source sequence x^n was emitted by the source. The encoder then looks into

---

[11]The smallest rate R for given distortion D is called the *rate distortion function*. The smallest distortion D for a given rate R is called the *distortion rate function*.

[12]$P_{\hat{X}X}$ can be determined from the chosen $P_{\hat{X}|X}$ and the given source distribution P_X.

its codebook and tries to find a codeword $\widehat{x}^n(u)$ such that

$$(x^n, \widehat{x}^n) \in T_\epsilon^n \left(P_{\widehat{X}X} \right) \tag{2.57}$$

If the encoder finds such an index it sends the corresponding index u to the decoder. The decoder receives u and puts out the reconstruction $\widehat{x}^n(u)$. The art is to find a good distribution $P_{\widehat{X}}$ such that the $\widehat{x}^n$ drawn according $P_{\widehat{X}}$ are good representations of x^n in the sense that the average distortion between X^n and $\widehat{X}^n$ is less than the specified average distortion D. By requiring that the source sequence x^n is jointly typical with the random code word $\widehat{x}^n$ we make sure that the empirical distribution of $\widehat{x}^n$ is close to $P_{\widehat{X}}$.

There are two sources of error in this approach. First, it may be that the source sequence is not typical, i.e., $x^n \notin T_\epsilon^n \left(P_X \right)$. However, from (2.4) we know that the probability of this event goes to zero as n grows to infinity. Second, suppose we have $x^n \in T_\epsilon^n \left(P_X \right)$, but none of the quantized $\widehat{x}^n(u)$, $u \in 1, 2, \ldots, 2^{nR}$ is jointly typical with the given x^n, i.e., $(x^n, \widehat{x}^n(u)) \notin T_\epsilon^n \left(P_{\widehat{X}X} \right)$ for all u[13]. The probability that none of the codewords $\widehat{x}^n(u)$ is suitable is upper bounded by

$$P_{\mathrm{e}}(x^n) = \left(1 - \Pr \left[(x^n, \widehat{X}^n) \in T_\epsilon^n \left(P_{\widehat{X}X} \right) \right] \right)^{2^{nR}} \tag{2.58}$$

$$\leq \left(1 - \left(1 - \epsilon_2'(n) \right) 2^{-n(I(X;\widehat{X})+\delta)} \right)^{2^{nR}} \tag{2.59}$$

$$\leq \exp \left(- \left(1 - \epsilon_2'(n) \right) 2^{n \left(R - I(X;\widehat{X}) - \delta \right)} \right) \tag{2.60}$$

where we used (2.49) to obtain the first inequality and $(1 - x)^n \leq \exp(-nx)$ for the second inequality. It follows that the rate has to satisfy

$$R > I(X; \widehat{X}) \tag{2.61}$$

in order to drive the probability $P_{\mathrm{e}}(x^n)$ to zero as n goes to infinity (note that $\lim_{n\to\infty} \epsilon_2'(n) = 0$). It remains to show that the average distortion between x^n and $\widehat{x}^n$ is less than D. Suppose $x^n \in T_\epsilon^n \left(P_X \right)$ and the decoder finds a $\widehat{x}^n(u)$ such that $(x^n, \widehat{x}^n(u)) \in T_\epsilon^n \left(P_{X\widehat{X}} \right)$. The distortion

---

[13]Remember: we want these sequences to be jointly typical because only then the joint empirical distribution of $(x, \widehat{x})$ is close to $P_{\widehat{X}X}$ and this is highly desirable since we designed the $P_{\widehat{X}X}$ (via $P_{\widehat{X}|X}$) to produce good (in the sense that the error is within D) quantized sequences.

follows as

$$D(x^n) = \frac{1}{n} \sum_{i=1}^{n} d(x_i, \hat{x}_i(u)) \tag{2.62}$$

$$= \frac{1}{n} \sum_{a,b} N(a, b | x^n, \hat{x}^n(u)) d(a, b) \tag{2.63}$$

$$\leq \sum_{a,b} \left(P_{X\hat{X}}(a, b) + \frac{\epsilon}{|\mathcal{X}||\hat{\mathcal{X}}|} \right) d(a, b) \tag{2.64}$$

$$= \sum_{a,b} P_{X\hat{X}}(a, b) d(a, b) + \epsilon \sum_{a,b} \frac{d(a, b)}{|\mathcal{X}||\hat{\mathcal{X}}|} \tag{2.65}$$

$$\leq \overline{D} + \epsilon d_{\max} \tag{2.66}$$

with $\overline{D} = \mathcal{E} \left\{ d(X, \hat{X}) \right\}$ and $d_{\max} = \max_{a,b} \{d(a, b)\}$. By increasing n the ϵ in (2.66) can be made as small as desirable and the distortion $D(x^n)$ is upper bounded by the average distortion $\overline{D}$. The rate distortion function is therefore given by

$$R(D) = \min_{P_{\hat{X}|X}: \, \mathcal{E}\left\{ d(X,\hat{X}) \right\} < D} I(X; \hat{X}), \tag{2.67}$$

i.e, choose the distribution $P_{\hat{X}|X}$ such that the rate becomes minimal and that the average distortion $\mathcal{E} \left\{ d(X, \hat{X}) \right\}$ is smaller than a specified distortion D. Note that

$$I(X; \hat{X}) = H(X) - H(X|\hat{X}), \tag{2.68}$$

i.e., in lossless source coding we required the rate R to be larger than $H(X)$, in lossy source coding we may further reduce the rate R by the amount of $H(X|\hat{X})$ by accepting some imperfection in the reconstruction.

Example: Gaussian Source

Consider a memoryless Gaussian source[14] with mean squared error distortion $\mathcal{E} \left\{ (X - \hat{X})^2 \right\} \leq D$. Let X be $\mathcal{N}(0, \sigma^2)$, where σ^2 is the variance of the source. By the rate distortion function (2.69) we have

$$R(D) = \min_{P_{\hat{X}|X}: \, \mathcal{E}\left\{ (X-\hat{X})^2 \right\} \leq D} I(X; \hat{X}). \tag{2.69}$$

---

[14]I.e., the alphabet of the source is continuous rather than discrete and each symbol emitted by the source is i.i.d. Gaussian.

From (2.68) we observe[15]

$$I(X; \widehat{X}) = h(X) - h(X|\widehat{X}) \tag{2.70}$$

$$= \frac{1}{2} \log \left(2\pi e\sigma^2\right) - h(X - \widehat{X}|\widehat{X}) \tag{2.71}$$

$$\geq \frac{1}{2} \log \left(2\pi e\sigma^2\right) - h(X - \widehat{X}) \tag{2.72}$$

$$\geq \frac{1}{2} \log \left(2\pi e\sigma^2\right) - \frac{1}{2} \log \left(2\pi e\mathcal{E}\left\{(X - \widehat{X})^2\right\}\right) \tag{2.73}$$

$$\geq \frac{1}{2} \log \left(2\pi e\sigma^2\right) - \frac{1}{2} \log \left(2\pi eD\right) \tag{2.74}$$

$$= \frac{1}{2} \log \left(\frac{\sigma^2}{D}\right). \tag{2.75}$$

The second term in the second equation follows because adding or subtracting $\widehat{X}$ to X does not change the entropy $h(X|\widehat{X})$, since $\widehat{X}$ is known. Inequality (2.72) follows from the fact that conditioning can only reduce entropy, inequality (2.73) is because the Gaussian distribution maximizes entropy when the second moment is given. Inequality (2.74) follows due to $\mathcal{E}\left\{(X - \widehat{X})^2\right\} \leq D$. We can conclude that

$$R(D) \geq \frac{1}{2} \log \left(\frac{\sigma^2}{D}\right). \tag{2.76}$$

The question arises whether the right hand side of (2.76) can be achieved. To find the conditional density $p(\widehat{x}|x)$ that achieves this lower bound it is more convenient to look at $p(x|\widehat{x})$. We choose for $D < \sigma^2$

$$X = \widehat{X} + Z \tag{2.77}$$

where $\widehat{X} \sim \mathcal{N}(0, \sigma^2 - D)$ and $Z \sim \mathcal{N}(0, D)$ are independent, i.e., $p(x|\widehat{x}) = p(z = x - \widehat{x})$. The mutual information $I(\widehat{X}; X) = I(X; \widehat{X})$ is then

$$I(X; \widehat{X}) = \frac{1}{2} \log \left(\frac{\sigma^2}{D}\right) \tag{2.78}$$

and the mean square error follows as $\mathcal{E}\left\{(X - \widehat{X})^2\right\} = D$, i.e., the lower bound (2.76) is achieved when the joint distribution $p(x, \widehat{x})$ is chosen according to (2.77). Note that for $D > \sigma^2$ we choose $\widehat{X} = 0$ with probability one and we get $R(D) = 0$, i.e., when the distortion D is chosen to be larger than the variance of the source σ^2, the distortion caused by Z drives the

---

[15]$h(X)$ denotes differential entropy of the continuous random variable X, see [74, Chapter 9].

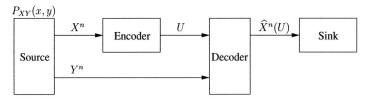

Fig. 2.5: The Wyner-Ziv problem or Lossy Source Coding With Side Information

mutual information between X and $\widehat{X}$ to zero.

We can rewrite (2.78) to obtain the distortion-rate function

$$D(R) = \sigma^2 2^{-2R}. \tag{2.79}$$

Each bit that is used to describe X reduces the distortion by a factor 4, i.e., each quantization bit reduces the quantization noise power by 6 dB.

2.4 Lossy Source Coding with Side Information or The Wyner-Ziv Problem

We have seen in the previous section that $R(D)$ bits are sufficient to describe a random variable within an average distortion D. Consider now a related problem depicted in Fig.2.5. A source emits a pair (X_i, Y_i) i.i.d. according to $P_{XY}(x, y)$ with $i = 1, 2, \ldots, n$. The encoder compresses the source sequence x^n and sends an index with rate R to the decoder. The decoder tries to determine an estimate $\widehat{x}^n$ from the index and the side information y^n, which is correlated with x^n and to which the decoder has access. The question is whether the encoder can reduce the number of bits $R(D)$ when it knows that the decoder has access to the sequence y^n. The answer is positive and shall be illustrated by following example.

Example. Assume a binary, discrete and memoryless source with alphabets $\mathcal{X} = \mathcal{Y} = \{0, 1\}$. Suppose a block length $n = 3$, i.e, the sequence X^3 in Fig.2.5 has eight possible outcomes after three clock times:

$$X^3 \in \mathcal{X}^3 = \{000, 001, 010, 011, 100, 101, 110, 111\}. \tag{2.80}$$

We assume that the sequence Y^3 has a Hamming distance to the sequence X^3 less or equal to one. If the decoder has no access to the Y^3 sequence, and the source emits a "zero" as often as a "one" in the X variable, the encoder cannot compress the X^3 sequence and has to use 3 bits for a lossless description[16]. If the encoder knows that the decoder has access to Y^3 and the encoder knows $P_{XY}(x, y)$, we can use following encoding:

- The encoder creates four bins and addresses them by two bits, $u^2 \in \mathcal{U}^2 = \{00, 01, 10, 11\}$.

- Each cell contains exactly two sequences $X^3 \in \mathcal{X}^3$.

- The sequences in each bin are chosen such, that the Hamming distance between both sequences is three, e.g., bin 00 contains the sequences $\{000, 111\}$, bin 01 contains $\{100, 011\}$, bin 10 contains $\{010, 101\}$ and bin 11 contains $\{001, 110\}$.

Note that this codebook construction is also known to the decoder. If the source emits for example the sequence $x^3 = 101$, the encoder instead transmits the corresponding bin index 10, using therefore a 2-bit description only. The decoder looks in his codebook at bin index 10 and has to find out, which of the two sequences $\{010, 101\}$ was originally emitted by the source. The decoder also knows $p_{XY}(x, y)$, i.e., in this specific example it knows that the Hamming distance between the y^3 sequence and the x^3 sequence has to be less or equal to one. The possible y^3 sequences for the emitted x^3 sequence 101 are $\{101, 111, 100, 001\}$. Assume for example that the observed y^3 is 100. As 100 has a Hamming distance two to the sequence 010 in bin 10 and a Hamming distance 1 to the sequence 101, the decoder puts out 101 as the reconstructed $\hat{x}^3$ sequence.

This example shows that side information at the decoder can help to reduce the number of bits necessary to describe a random variable[17]. We extend now the rate distortion problem discussed in the previous section to the case where side information is available at the decoder side.

We introduce an auxiliary random variable Q with distribution $P_{Q|X}(q|x)$ that is chosen freely, where $Q \in \mathcal{Q}^{18}$. Q^n represents a codeword of length n representing X^n and is sent to the decoder. The decoder observes the sequences q^n (codeword) and y^n (side information) and puts out the reconstruction $\hat{x}^n = f(q^n, y^n)$.

Codebook construction. Generate $2^{n(R+R')}$ codewords $q^n(w, v)$, $w = 1, 2, \ldots, 2^{nR}$, $v = 1, 2, \ldots, 2^{nR'}$ by choosing the $n2^{n(R+R')}$ symbols i.i.d. according to $P_Q(\cdot)$, which is computed

---

[16]In this example we look at lossless descriptions in order to simplify the motivation why to use side information in source coding.

[17]Of course the side information has to be correlated with the random variable to be described/encoded.

[18]In most cases the quantization alphabet $\mathcal{Q}$ is chosen to be equal to the source alphabet $\mathcal{X}$.

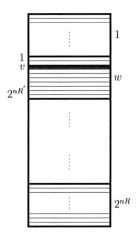

Fig. 2.6: Codebook construction for the Wyner-Ziv Problem. w addresses the bins and v addresses the codewords within one bin. The shaded row represents a codeword $q^n(w, v)$.

from $P_{XQ(\cdot,\cdot)}$. The code book contains 2^{nR} bins, addressed by the index w and $2^{nR'}$ codewords $q^n(w, v)$ within each bin, addressed by the relative index v. Fig.2.6 illustrates the code construction.

Encoder. Given a sequence x^n, the encoder tries to find a pair (w, v) such that

$$(x^n, q^n(w, v)) \in T_\epsilon^n (P_{XQ}). \tag{2.81}$$

If the encoder finds such a codeword it sends the bin index w to the decoder.

Decoder. Given the bin index w and the side information y^n, the decoder tries to find an index v such that

$$(q^n(w, v), y^n) \in T_\epsilon^n (P_{QY}) \tag{2.82}$$

If there is such a unique v, then the decoder puts out $\hat{x}^n = q^n(w, v)$ as the reconstructed sequence. If there are several candidates that satisfy the typicality check, then the decoder chooses one of them at random and puts out the corresponding reconstruction $\hat{x}^n$.

We see that instead of sending the complete position of the codeword[19], i.e., the index pair (w, v), we send only the index of the bin where the codeword lies, which is w. If the number of codewords in each bin is small enough, then the side information y^n at the decoder can be used

---

[19]As it is done in rate distortion theory without side information.

to isolate the particular codeword in that bin. The question now is how to design the rates R and R', such that R is as small as possible and the average reconstruction distortion is less or equal than a fixed value D.

Analysis.

a) Suppose that $(x^n, y^n) \notin T_\epsilon^n (P_{XY})$. According to (2.4) (extended to joint typicality), the probability of this event goes to zero as n grows large.

b) Suppose that $(x^n, y^n) \in T_\epsilon^n (P_{XY})$, but the encoder cannot find a (w, v) such that $(x^n, q^n(w, v)) \in T_\epsilon^n (P_{XQ})$. From rate distortion theory, i.e., (2.61), we know that this probability is small if

$$R + R' > I(X; Q) \tag{2.83}$$

c) Suppose the pair of sequences (x^n, y^n) is jointly typical and the encoder finds a (w, v) such that $(x^n, q^n(w, v))$ is jointly typical. However, the decoder cannot find a v such that $(q^n(w, v), y^n) \in T_\epsilon^n (P_{QY})$, i.e., the encoder cannot find a codeword that is jointly typical with the y^n sequence. By the Markov Lemma 2.1.5 the probability of this event is small. When the source sequence x^n is jointly typical with the side information y^n and the source sequence x^n is jointly typical with the codeword $q^n(w, v)$, then we also must have that y^n is jointly typical with the codeword $q^n(w, v)$, since by the code construction the variables Y–X–Q form a Markov chain[20].

c) Suppose the pair of sequences (x^n, y^n) is jointly typical and the encoder finds a (w, v) such that $(x^n, q^n(w, v))$ is jointly typical. However, the decoder finds another $\tilde{v} \neq v$ such that $(q^n(w, \tilde{v}), y^n)$, i.e., there are two or more codewords within the same bin that are jointly typical with the side information sequence y^n. Since the probability that a randomly chosen q^n is jointly typical with y^n is upper bounded by Theorem 2.1.3, the probability that there is another $q^n(w, \tilde{v})$ in the same bin that is jointly typical with Y^n is upper bounded by the number of codewords within the bin (that is $2^{nR'}$) times the probability of joint typicality, i.e.,

$$\Pr\left[\exists \tilde{v} \neq v : (q^n(w, \tilde{v}), y^n) \in T_\epsilon^n (P_{QY})\right] \leq 2^{nR'} 2^{-n(I(Q;Y) - \delta)} \tag{2.84}$$

which goes to zero as $R' < I(Y; Q) - \delta$ and n grows large.

d) It can be shown [97] along the same lines as in the case with no side information that the

---

[20]Note that the joint distribution can be written as $P_{YXQ} = P_Y P_{X|Y} P_{Q|XY} = P_Y P_{X|Y} P_{Q|X}$, where $P_Y P_{X|Y}$ is given by the source and $P_{Q|X}$ is chosen freely and is a design "parameter".

average distortion between x^n and $\hat{x}^n$ is less than D when all typicality properties have been satisfied.

Combining (2.83) and (2.84) we obtain the achievable rate of the Wyner-Ziv source coding problem

$$R(D) = \min_{P_{Q|X}:\, \mathcal{E}\{d(X,\hat{X})\}\leq D} I(X;Q) - I(Y;Q) = I(X;Q|Y), \qquad (2.85)$$

where the last equality was obtained by

$$I(X;Q) - I(Y;Q) = H(Q) - H(Q|X) - H(Q) + H(Q|Y) \qquad (2.86)$$

$$= H(Q|Y) - H(Q|X) \qquad (2.87)$$

$$= H(Q|Y) - H(Q|X,Y) \qquad (2.88)$$

$$= I(X;Q|Y). \qquad (2.89)$$

The third equality is because the variable Y depends only on X but not on Q, and therefore conditioning on Y does not change the entropy of Q. The mutual information $I(X;Q|Y)$ can be interpreted as how many information is contained in the quantized codeword Q^n about the source sequence X^n, assuming that Y^n (which is correlated with X^n) is known at the decoder.

Example: Gaussian Source

Suppose that X and Y are zero mean Gaussian random variables with variances σ_X^2 and σ_Y^2, respectively, and with correlation coefficient $\rho = \mathcal{E}\{XY\}/(\sigma_X\sigma_Y)$. The distortion is defined as

$$\mathcal{E}\left\{(X - \hat{X})^2\right\} \leq D. \qquad (2.90)$$

As in the example of lossy source coding with no side information we choose the auxiliary random variable $Q = X + Z$, where Z is Gaussian distributed with zero mean and variance σ_Z^2. The decoder puts out the reconstruction

$$\hat{X} = f(Y,Q) = \mathcal{E}\{X|Y,Q\} \qquad (2.91)$$

which is the minimum mean square estimator (MMSE) of X given Y and Q. We can now compute (2.89)

$$I(X;Q|Y) = h(X|Y) - h(X|YQ) \tag{2.92}$$

$$= h(X|Y) - h(X - \widehat{X}|YQ) \tag{2.93}$$

$$= h(X|Y) - h(X - \widehat{X}) \tag{2.94}$$

$$R(D) = \frac{1}{2}\log\left(\frac{\sigma_X^2(1 - \rho^2)}{D}\right) \tag{2.95}$$

where the second equality follows since $\widehat{X}$ is a deterministic function of Y and Q and therefore adding or subtracting $\widehat{X}$ to X does not change the entropy for given Y and Q. The third equality follows due to the orthogonality property of the MMSE estimation, i.e, the estimation error $X - \widehat{X}$ is orthogonal (uncorrelated) to the observations Y and Q and therefore, conditioning on Y and Q has no influence on the entropy of the estimation error. The last equality follows by the same reasons as in the Gaussian example in the previous section where no side information was available. Note that if we choose $D > \sigma_X^2(1 - \rho^2)$, we have $R(D) = 0$, since the disturbance of X by Z is larger than the variance of the unknown part of the source. For $\rho > 0$ the rate in (2.95) is smaller than the rate in (2.78), i.e., the side information Y helps to reduce the source coding rate. If $\rho = 0$, both rates are the same, i.e., side information Y is useless, since it is uncorrelated with X.

Wyner-Ziv coding will be applied in the next two chapters to relay channels. We will see that in the compress-and-forward strategy originally proposed in [14] the relay acts as a source encoder and quantizes its receive signal before transmitting it to the destination. The destination's receive signal serves as side information (since both the relay's and the destination's receive signals are influenced by the transmitted signal of the source and are therefore correlated) in the decoding process at the destination.

The next chapter reviews the capacity results of full-duplex and half-duplex relay channels, where we will make use of the concepts introduced in this chapter.

3 Capacity Theorems for the Relay Channel

This chapter introduces the *relay channel*. In Section 3.1 we review the capacity results obtained by Cover and Gamal in the late seventies [14]. The capacity of the general relay channel is still unknown, however, Cover and Gamal provided an upper bound and several lower bounds on the capacity of the relay channel. They also specialized their bounds to the case of the AWGN relay channel. The upper bound is based on a max-flow min-cut theorem (cut-set bound) [74] and defines an ultimate limit on the communication rate. It is still not known but commonly assumed that this upper bound is not tight in general (only in certain cases). The first lower bound, today known as *decode-and-forward*, is based on superposition coding and enables the source and the relay to cooperate such, that the received symbols at the destination add up coherently. The second lower bound, today known as *compress-and-forward*, makes use of lossy source coding with side information [98] that was introduced in the previous chapter. In Section 3.2 we discuss the application of these methods to fading AWGN relay channels for the case of full-duplex and half-duplex relay devices. We review recent results obtained by Kramer et al. [15], [99], Madsen and Zhang [16], and Khojastepour et al. [77]. The main result is that the cut-set upper bound can be achieved by decode-and-forward when the relay is in the neighborhood of the source terminal. When the relay is near to the destination terminal the cut-set bound is achieved by the compress-and-forward strategy. Hence, these results show that the capacity for wireless relay channels is known in certain cases.

3.1 Capacity Theorems for the Relay Channel

3.1.1 System Model

In this section we consider the discrete memoryless relay channel depicted in Fig.3.1 with terminal T_1 being the source, terminal T_2 the destination and terminal T_3 the relay.

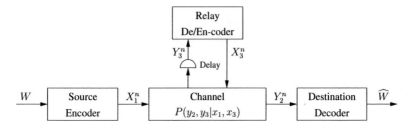

Fig. 3.1: The relay channel

The random variables of the relay channel are:

- the message $W \in \mathcal{W}$ of the source terminal T_1,

- the channel inputs (transmitted symbols) $X_{ti} \in \mathcal{X}_t$, $t = 1, 3$, $i = 1, 2, \ldots, n$,

- the channel outputs (received symbols) $Y_{ti} \in \mathcal{Y}_t$, $t = 2, 3$, $i = 1, 2, \ldots, n$,

- the message estimate $\widehat{W} \in \mathcal{W}$ at the destination terminal T_2,

where $\mathcal{W} = \{1, 2, \ldots, 2^{B_W}\}$ is the message set. The message W to be transmitted is chosen randomly from $\mathcal{W}$ according to a uniform distribution and carries B_W information bits. The discrete input alphabets are given by $\mathcal{X}_1$, $\mathcal{X}_3$ and the discrete output alphabets by $\mathcal{Y}_2$, $\mathcal{Y}_3$. The length of each transmitted codeword is denoted by n. Codeword $X_1^n = (X_{11}, X_{12}, \ldots, X_{1n})$ of source terminal T_1 is a function of the message W

$$f : \mathcal{W} \to \mathcal{X}_1^n \qquad (3.1)$$

yielding codewords $X_1^n(1), X_1^n(2), \ldots, X_1^n\left(2^{B_W}\right)$, where $\mathcal{X}_1^n$ denotes the set of all length n source sequences with symbol alphabet $\mathcal{X}_1$. The transmit symbol X_{3i} of relay terminal T_3 is a function of the relay's past received symbols $Y_3^{i-1} = (Y_{31}, Y_{32}, \ldots, Y_{3i-1})$, i.e.,

$$X_{3i} = f_i(Y_{31}, Y_{32}, \ldots, Y_{3i-1}), \qquad i \leq n. \qquad (3.2)$$

Relation (3.2) follows from the assumption that the relay operates in a causal fashion. In [100] a different model is considered, where each transmitted relay symbol can depend on the relay's current as well as past received symbols (*relay-without-delay*). Such a model is appropriate if the processing delay at the relay is small compared to the delay of the source to destination transmission.

Destination terminal T_2 computes its message estimate $\widehat{W}$ as a function of its received symbols $Y_2^n = (Y_{21}, Y_{22}, \ldots, Y_{2n})$. Therefore, the decoding function at the destination terminal T_2 is defined as

$$g : \mathcal{Y}_2^n \to \mathcal{W}. \tag{3.3}$$

We assume a time invariant and memoryless relay channel which is defined probabilistically by the conditional channel distribution

$$P_{Y_2 Y_3 | X_1 X_3}(y_2, y_3 | x_1, x_3) \tag{3.4}$$

where X_1, X_3 and Y_2, Y_3 are random variables representing the respective channel inputs and outputs. The joint probability distribution[1] of all random variables factors then as [101]

$$P(w)P(x_1^n|w)\left[\prod_{i=1}^{n} P(x_{3i}|y_3^{i-1})P(y_{2i},y_{3i}|x_{1i},x_{3i})\right]P(\widehat{w}|y_2^n) \tag{3.5}$$

where $P(x_1^n|w)$, $P(x_{3i}|y_3^{i-1})$ and $P(\widehat{w}|y_2^n)$ take on the values 0 and 1 only. To see this, consider for example $P(x_1^n|w)$: there are $|\mathcal{X}_1|^n$ possible sequences, but to map the message W to a codeword we only need 2^{B_W} sequences to form a codebook. The probability $P(x_1^n|w)$ takes on the value 0 for all sequences x_1^n not belonging to the chosen codebook and takes on the value 1 for the sequences belonging to the chosen codebook.

The positive number R is called an achievable rate if for any $\delta > 0$, $\epsilon > 0$ and sufficiently large n there exist encoding and decoding functions such that $\frac{B_W}{n} \geq R - \delta$ with $\Pr(\widehat{W} \neq W) \leq \epsilon$. The capacity C of the relay channel is the supremum of all achievable rates [74, Chapter 8].

3.1.2 Cut-set Upper Bound

An upper bound on the capacity of the relay channel can be given by applying the cut-set bound to the relay network [74]. For a general multi-terminal network as illustrated in Fig.3.2 the maximum achievable sum-rate across any cut-set is upper bounded by

$$\sum_{i \in \mathcal{S}, j \in \mathcal{S}^c} R^{(ij)} \leq I(X^{(\mathcal{S})}; Y^{(\mathcal{S}^c)} \mid X^{(\mathcal{S}^c)}) \tag{3.6}$$

---

[1]From now on subscripts in the probability distributions are dropped when they are obvious by inspection of the arguments.

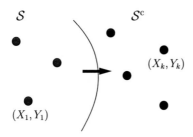

Fig. 3.2: Cut-set in a general multi-terminal network

where $R^{(ij)}$ is the rate from terminal $i \in \mathcal{S}$ to terminal $j \in \mathcal{S}^c$. Then $X^{(\mathcal{S})}$ denotes the transmit symbols of all terminals in $\mathcal{S}$, $X^{(\mathcal{S}^c)}$ the transmit symbols of all terminals in $\mathcal{S}^c$ and $Y^{(\mathcal{S}^c)}$ the received symbols of all terminals in $\mathcal{S}^c$. The proof of (3.6) is given in [74, Theorem 14.10.1]. The cut-set bound states that the rate of information flow across any boundary is upper bounded by the mutual information between the transmit symbols one side and the received symbols on the other side, conditioned on the transmit symbols on the other side. The bound is equal to the rate achievable in a vector channel, i.e., where all transmitting terminals in $\mathcal{S}$ can cooperate arbitrarily and all receiving terminals in $\mathcal{S}^c$ can cooperate arbitrarily. Of course, such a rate can only be larger than a rate, where the terminals are not allowed or only partially allowed to cooperate. The conditioning in (3.6) says, that terminals in $\mathcal{S}^c$ are also allowed to transmit during they receive messages from terminals[2] in $\mathcal{S}$, and that the $X^{(\mathcal{S}^c)}$ can be chosen jointly by all nodes[3].

This bound can now be applied to the relay network with different choices of cut-sets, see Fig.3.3. The *broadcast cut* upper bounds the maximum rate from the source to both relay and destination and is given by

$$R_{\text{BC}} \leq I(X_1; Y_2 Y_3 | X_3). \tag{3.7}$$

In order to achieve this bound the relay and the destination must decode the codeword of the source jointly, i.e., practically, the antennas of the relay and the destination have to be co-located. The *multiple access cut* upper bounds the maximum rate from both the source and the relay to the destination and is given by

$$R_{\text{MA}} \leq I(X_1 X_3; Y_2). \tag{3.8}$$

---

[2]This requires full-duplex terminals.

[3]Looking at the relay channel we will see, why a node in $\mathcal{S}^c$ should transmit at all. It turns out, that for example terminal i in $\mathcal{S}^c$ can help terminal j in $\mathcal{S}^c$ to increase the data rate of terminal j by transmitting an appropriate $x_i^{(\mathcal{S}^c)}$.

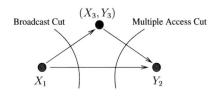

Fig. 3.3: Cut-sets for a relay network

This bound can be achieved when the source and the relay can jointly encode the message to be transmitted, i.e., practically, the antennas of the source and the relay have to be co-located. Note that in (3.8) there is no conditioning on a transmit symbol because the destination does only receive symbols and does not transmit any own symbols. An achievable rate of the relay channel is then upper bounded by the minimum of (3.7) and (3.8). The maximum rate (capacity) is upper bounded by maximizing the minimum of (3.7) and (3.8) over all probability distributions of the channel inputs X_1 (source) and X_3 (relay):

Proposition 3.1.1 (Capacity Upper Bound for the Relay Channel).

$$C \leq \max_{P(x_1, x_3)} \min \Big(I(X_1; Y_2 Y_3 | X_3), I(X_1 X_3; Y_2) \Big). \tag{3.9}$$

Note that for the broadcast cut the relay and the destination should perfectly cooperate in order to achieve this bound, whereas for the multiple access cut the source and the relay should perfectly cooperate to achieve the upper bound. Since both levels of cooperation cannot be achieved simultaneously it is believed (but not proved) that the cut-set bound (3.9) is not tight for the general relay channel. However, it was shown in [15] and [16] that for wireless relay channels the cut-set upper bound can be achieved when the relay is near to the source or near to the destination.

Next we describe two schemes that establish lower bounds on the capacity of the relay channel. The first scheme, decode-and-forward, achieves the cut-set bound on the capacity when the multiple access cut is the bottleneck in (3.9). The second scheme, compress-and-forward, achieves the cut-set bound on the capacity when the broadcast cut is the bottleneck in (3.9).

3.1.3 Decode-and-forward

The idea of this scheme is to let the relay and the source cooperate such, that the transmitted symbols of the source and the relay are correlated. This leads to a coherent addition at the

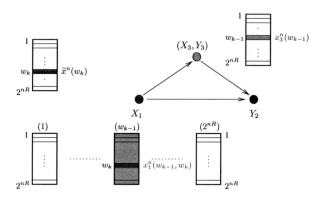

Fig. 3.4: Code book construction for the decode-and-forward relay channel. There are 2^{nR} + 1 different codebooks at the source. To each of the 2^{nR} code *books* at the source corresponds one code*word* at the relay. w_{k-1} determines the codeword at the relay and says, which code book has to be used at the source. w_k then determines, which codeword in this code book is going to be transmitted.

destination and therefore to an SNR (signal-to-noise ratio) gain. We describe now how the source and the relay cooperate to achieve this SNR gain. The scheme is based on block Markov superposition coding with sliding window decoding at the destination [14], [101].

Block Markov Superposition Coding. At the source terminal T_1 the message w is divided into K blocks $w_1, w_2, \ldots, w_K$ of nR bits each. The transmission is performed in $K + 1$ blocks, i.e., $B_W = KnR$ message bits are transmitted in $(K + 1)n$ channel uses with an overall rate $RK/(K + 1)$ which is for $K \to \infty$ and for fixed n arbitrary close to R. The codeword of length n at terminal T_1 consists of a superposition of two codewords, see Fig. 3.4:

$$x_1^n(i, j) = \sqrt{\alpha} x_3^n(i) + \sqrt{\beta} \widetilde{x}_1^n(j) \tag{3.10}$$

where i and j range from 1 to 2^{nR} and α, β are scaling coefficients that will be explained later. The $n \cdot 2^{nR}$ symbols $\widetilde{x}_{1m}(j)$ are chosen independently according to $P_{\widetilde{X}_1}(\cdot)$ and the $n \cdot 2^{nR}$ symbols $x_{3m}(i)$ are chosen independently according to $P_{X_3}(\cdot)$. In the first block, $k = 1$, terminal T_1 transmits $x_1^n(1, w_1)$. The relay terminal T_3 transmits $x_3^n(1)$, since at that moment it has no knowledge about the message sent by the source. After block 1 the relay is able to decode w_1 as long as

$$R \leq I(X_1; Y_3 | X_3) \tag{3.11}$$

$x_1(1, w_1)$	$x_1(w_1, w_2)$	$x_1(w_2, w_3)$	$x_1(w_3, 1)$
$x_3(1)$	$x_3(w_1)$	$x_3(w_2)$	$x_3(w_3)$

Fig. 3.5: Block Markov encoding for $K = 4$

and the block size n is large. In the second block, $k = 2$, terminal T_1 transmits $x_1^n(w_1, w_2)$ and the relay terminal transmits $x_3^n(w_1)$. Relay T_3 is able to decode w_2 as long as n is large and (3.11) is true. One continues in this way until block $K + 1$, where terminal T_1 transmits $x_1^n(w_K, 1)$. The transmission scheme is depicted in Fig.3.5. Note that the conditioning on X_3 in (3.11) means that the relay is a full-duplex relay and transmits X_3 during receiving Y_3 but the influence of X_3 on Y_3 is known (and can be for example canceled).

Sliding Window Decoding. Let y_{2k}^n be the received symbols at the destination terminal T_2 in block k and $y_{2(k-1)}^n$ the received symbols in block $k - 1$. Starting after the second block $k = 2$, terminal T_2 decodes w_1 from y_{22}^n and y_{21}^n as long as

$$R \leq I(X_3; Y_2) + I(X_1; Y_2|X_3) \tag{3.12}$$
$$= I(X_1 X_3; Y_2) \tag{3.13}$$

and the block size n is large. (3.13) follows by the chain rule of mutual information [74, Theorem 2.5.2]. The contribution $I(X_3; Y_2)$ is from y_{22}^n and the contribution $I(X_1; Y_2|X_3)$ is from y_{21}^n. After decoding w_1 the decoder shifts the window by one block and decodes w_2 from y_{23}^n and y_{22}^n. Since w_1 is already decoded, it does not act as interference in y_{22}^n. The message w_2 can be decoded again as long as (3.13) is fulfilled and n is large. One continues this way until block $k = K$ where finally all messages $w_1, \ldots, w_K$ are decoded. Combining (3.11) and (3.12) we obtain the achievable rate for this scheme in the following proposition.

Proposition 3.1.2 (Achievable Rate for Decode-and-forward).

$$R = \max_{P(x_1, x_3)} \min \left(I(X_1; Y_3|X_3), I(X_1 X_3; Y_2) \right). \tag{3.14}$$

For the proof of the proposition one has to show that as long as the rate is less or equal than (3.14) the probability of decoding error can be made arbitrary small. We omit the proof here and refer to [14] or [101]. The rate given in (3.14) can be approached by different strategies:

- irregular encoding/successive decoding, [14]

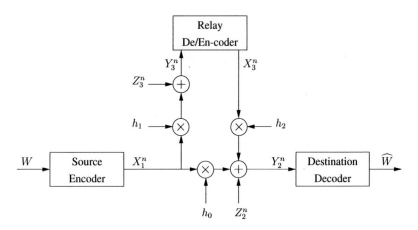

Fig. 3.6: The AWGN relay channel

- regular encoding/sliding window decoding, [15]

- regular encoding/backward decoding [97]

where regular encoding with sliding window decoding is the coding scheme described in this section. See [15] for a discussion of these methods.

Consider next the AWGN channel of Fig.3.6 with inputs and outputs taken from the complex alphabet $\mathbb{C}$. The input-output relations for the AWGN relay channel are defined by

$$Y_2 = X_1 h_0 + X_3 h_2 + Z_2 \tag{3.15}$$

$$Y_3 = X_1 h_1 + Z_3 \tag{3.16}$$

where $h_1 \in \mathbb{C}$ is the deterministic channel gain between source and relay, $h_2 \in \mathbb{C}$ the deterministic channel gain between relay and destination and $h_0 \in \mathbb{C}$ the deterministic channel gain between source and destination. The noise random variable at the destination is denoted by $Z_2 \sim \mathcal{CN}(0, \sigma_2^2)$ and the noise random variable at the relay is denoted by $Z_3 \sim \mathcal{CN}(0, \sigma_3^2)$ and both variables are independent of each other and independent of the inputs X_1 and X_3. Note that the influence of X_3 on the relay reception (self-interference) is assumed to be perfectly canceled at the relay, therefore X_3 does not appear in (3.16). The codeword at the source terminal is chosen as

$$x_1^n(i,j) = \sqrt{\alpha} x_3^n(i) + \sqrt{\beta} \tilde{x}_1^n(j) \tag{3.17}$$

with i.i.d. $\widetilde{X}_{1l} \sim \mathcal{CN}(0, P_1)$ and i.i.d. $X_{3l} \sim \mathcal{CN}(0, P_3)$ for $l = 1, 2, \ldots, n$. The scaling coefficients are chosen such, that the average power of the source codeword is equal to P_1, i.e.,

$$\frac{1}{n} \sum_{l=1}^{n} |x_{1l}(i, j)|^2 = P_1 \tag{3.18}$$

for all integers $i, j \in [1, 2^{nR}]$. From that the scaling coefficients follow as

$$\alpha = \frac{|\rho|^2 P_1}{P_3} \tag{3.19}$$

$$\beta = (1 - |\rho|^2). \tag{3.20}$$

The parameter ρ denotes the complex correlation coefficient between source symbol X_1 and relay symbol X_3 and is defined as

$$\rho = \frac{\mathcal{E}\{X_1 X_3^*\}}{\sqrt{P_1 P_3}}. \tag{3.21}$$

The mutual information between source and relay (3.11) can now be evaluated as

$$I(X_1; Y_3 | X_3) = h(Y_3 | X_3) - h(Y_3 | X_1, X_3) \tag{3.22}$$

$$= h(Y_3 | X_3) - h(Z_3) \tag{3.23}$$

$$= h(Y_3, X_3) - h(X_3) - h(Z_3) \tag{3.24}$$

$$= \log\left((\pi e)^2 |R_{Y_3 X_3}|^2\right) - \log\left(\pi e P_3\right) - \log\left(\pi e \sigma_3^3\right). \tag{3.25}$$

The covariance matrix $R_{Y_3 X_3}$ follows as

$$R_{Y_3 X_3} = \begin{pmatrix} P_1 |h_1|^2 + \sigma_3^2 & h_1 \mathcal{E}\{X_1 X_3^*\} \\ h_1^* \mathcal{E}\{X_1^* X_3\} & P_3 \end{pmatrix}. \tag{3.26}$$

Inserting the determinant of (3.26) into (3.25) leads to

$$I(X_1; Y_3 | X_3) = \log\left(\pi e \left(P_1 |h_1|^2 \left(1 - |\rho|^2\right) + \sigma_3^2\right)\right) - \log\left(\pi e \sigma_3^2\right) \tag{3.27}$$

$$= \log\left(1 + \frac{P_1 |h_1|^2 \left(1 - |\rho|^2\right)}{\sigma_3^3}\right) \tag{3.28}$$

where we used the fact that $\mathcal{E}\{X_1^* X_3\} = \mathcal{E}\{(X_1 X_3^*)^*\} = \mathcal{E}\{X_1 X_3^*\}^* = \sqrt{P_1 P_3} \rho^*$. The mutual information of the multiple access cut (3.12) follows as

$$I(X_1X_3; Y_2) = h(Y_2) - h(Y_2|X_1, X_3) \tag{3.29}$$

$$= h(X_1h_0 + X_3h_2 + Z_2) - h(Z_2) \tag{3.30}$$

$$= \log\left(\pi e \cdot \mathcal{V}\left\{X_1h_0 + X_3h_2 + Z_2\right\}\right) - \log\left(\pi e \sigma_2^2\right) \tag{3.31}$$

$$= \log\left(1 + \frac{P_1|h_0|^2 + P_3|h_2|^2 + 2\Re\left\{h_0h_2^*\rho\right\}\sqrt{P_1P_3}}{\sigma_2^2}\right) \tag{3.32}$$

where $\mathcal{V}\{\cdot\}$ denotes the variance operator. Summarizing the results we obtain the achievable rate of the decode-and-forward scheme for the AWGN relay channel:

Proposition 3.1.3 (Achievable Rate for AWGN Decode-and-forward).

$$R = \max_{0 \leq \rho_r \leq 1} \min\left(C\left(\frac{P_1|h_1|^2(1-\rho_r^2)}{\sigma_3^2}\right), \right.$$
$$\left. C\left(\frac{P_1|h_0|^2 + P_3|h_2|^2 + 2\rho_r\sqrt{P_1P_3}|h_0||h_2|\cos\left(\phi_0 - \phi_2\right)}{\sigma_2^2}\right)\right). \tag{3.33}$$

where

$$\rho = \Re\left\{\rho\right\} + j\Im\left\{\rho\right\} = \rho_r + j\rho_i, \tag{3.34}$$

$C(x) = \log(1 + x)$, ϕ_1 and ϕ_2 the phases of the channels h_1 and h_2, respectively. For the maximum rate, one has to choose $\rho_i^2 = 0$, since the imaginary part does not influence the second term in (3.33) and can only decrease the source-relay mutual information. The optimization has therefore to be done only over the real part of the complex correlation coefficient given in (3.21). The range of ρ_r is by definition $[-1, 1]$, but negative values can be excluded since they can only decrease the second term in (3.33), whereas the sign of ρ_r does not influence the source-relay mutual information. The first term in (3.33) denotes the rate at which the relay can decode the messages transmitted by the source. Note that the available power for this task is reduced by the factor $(1 - \rho_r^2)$ since X_1 and X_3 are correlated and $(1 - \rho_r)^2P_1$ denotes the variance of the unknown amount of information at the relay, i.e., only $(1 - \rho_r^2)P_1$ is used to transmit the new information to the relay and the remaining power is spent for the cooperation between source and relay. The second term describes the rate that can be achieved when the source and the relay jointly transmit correlated symbols to the destination and therefore mimics a two-input one-output channel.

In order to compare the decode-and-forward rate with the cut-set upper bound (3.9) we also evaluate the bound for the AWGN case. The only difference between (3.1.2) and (3.9) lies in the first term of the minimization, i.e., instead of a source-relay cut there is a broadcast cut:

$$I(X_1; Y_2Y_2|X_3) = h(Y_2, Y_3|X_3) - h(Y_2Y_3|X_1X_3) \tag{3.35}$$

$$= h(Y_2, Y_3, X_3) - h(X_3) - h(Z_2) - h(Z_3) \tag{3.36}$$

$$= \log\left((\pi e)^3 |R_{Y_3Y_3X_3}|\right) - \log\left(\pi e P_3\right)$$

$$\quad - \log\left(\pi e \sigma_2^2\right) - \log\left(\pi e \sigma_3^2\right) \tag{3.37}$$

$$= \log\left(1 + P_1(1 - \rho_r^2)\left(\frac{|h_0|^2}{\sigma_2^2} + \frac{|h_1|^2}{\sigma_3^2}\right)\right) \tag{3.38}$$

where the covariance matrix $R_{Y_2Y_3X_3}$ is similarly determined as in (3.26). We obtain now:

Proposition 3.1.4 (Cut-set Upper Bound for AWGN Relay Channel).

$$C \leq \max_{0 \leq \rho_r \leq 1} \min\left(C\left(P_1(1 - \rho_r^2)\left(\frac{|h_0|^2}{\sigma_2^2} + \frac{|h_1|^2}{\sigma_3^2}\right)\right),\right.$$
$$\left. C\left(\frac{P_1|h_0|^2 + P_3|h_2|^2 + 2\rho_r\sqrt{P_1P_3}|h_0||h_2|\cos\left(\phi_0 - \phi_2\right)}{\sigma_2^2}\right)\right). \tag{3.39}$$

As mentioned, the expressions (3.33) and (3.39) differ only in the first term. In the decode-and-forward scheme the relay is required to fully decode the message transmitted by the source whereas in the cut-set upper bound this is not required. We see that the DF scheme approaches the cut-set upper bound when the second term in the cut-set bound is the bottleneck, since then the expressions (3.33) and (3.39) are identical. Numerical examples are provided in Section 4.6.

3.1.4 Compress-and-forward

In the DF transmission scheme the relay was enforced to decode the symbols from the source. When the source-relay channel is weak (e.g., due to a large noise variance at the relay and/or a weak channel gain between source and relay) the rate achievable from source to destination is limited by the source-relay capacity. In [14, Theorem 6] the authors proposed a second relaying strategy that leads to a second lower bound. In this strategy, nowadays known as *compress-and-forward*[4], the relay does not try to decode the message, but sends estimates of the received symbols to the destination. The achievable rate with this scheme is given now.

Proposition 3.1.5 (Achievable Rate for Compress-and-forward). *An achievable rate for the*

---

[4]In the literature this strategy is sometimes called *quantize-and-forward* or *estimate-and-forward*.

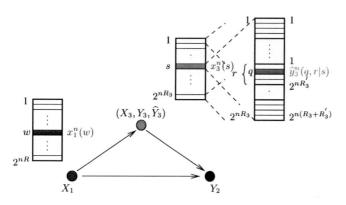

Fig. 3.7: Code book constructions for compress-and-forward

relay channel is given by

$$R \leq \max_{P(x_1)P(x_3)P(\widehat{y}_3|x_3,y_3)} I(X_1; Y_2\widehat{Y}_3|X_3) \tag{3.40}$$

subject to the constraint

$$I(\widehat{Y}_3; Y_3|X_3Y_2) \leq I(X_3; Y_2). \tag{3.41}$$

The auxiliary random variable $\widehat{Y}_3$ represents a quantized and compressed version of Y_3. For a received Y_3, the relay chooses a $\widehat{Y}_3$ according to its quantization rule, encodes $\widehat{Y}_3$ using a channel code and transmits the index of the chosen $\widehat{Y}_3$ to the destination. Due to the constraint in (3.41) the destination can reliably decode the transmitted index and recover $\widehat{Y}_3$. The destination can now use Y_2 and $\widehat{Y}_3$ to decode the message from the source, mimicking a one-input two-output channel. We see that this corresponds to a Wyner-Ziv problem, with the difference that the the bin index of the quantized codeword is not directly available at the decoder, but has to be again encoded (using a channel code) and to be sent over the relay-destination channel to the destination decoder. At the same time the destination also receives symbols directly from the source and these symbols are used as side information to reconstruct $\widehat{Y}_3$. In the Wyner-Ziv setting, Y_3 plays the role of the "source sequence".

Outline of Proof. The proof is based on the use of *typical sequences*[5]. The full proof can be found in [14] or [101]. The next chapter will provide the proof of the compress-and-forward scheme for the two-way relay channel which is similar to the proof of the one-way relay channel.

---

[5]An introduction into the theory of typical sequences was given in Chapter 2.

Construction of Code books. The source terminal T_1 chooses 2^{nR} i.i.d. codewords x_1^n each with probability

$$P^n(x_1^n) = \prod_{i=1}^{n} P(x_{1i}) \tag{3.42}$$

and labels the codewords by $w \in [1, 2^{nR}]$, see Fig.3.7. The relay terminal T_3 chooses 2^{nR_3} i.i.d. codewords x_3^n each with probability

$$P^n(x_3^n) = \prod_{i=1}^{n} P(x_{3i}) \tag{3.43}$$

and labels the codewords by $s \in [1, 2^{nR_3}]$. The relay chooses then, for each $x_3^n(s)$, $2^{n(R_3+R_3')}$ i.i.d. $\widehat{y}_3^n$ each with probability

$$P^n(\widehat{y}_3^n | x_3^n(s)) = \prod_{i=1}^{n} P(\widehat{y}_{3i} | x_{3i}(s)) \tag{3.44}$$

with

$$P(\widehat{y}_3 | x_3) = \sum_{x_1, y_2, y_3} P(x_1) P(y_2, y_3 | x_1, x_3) P(\widehat{y}_3 | y_3, x_3) \tag{3.45}$$

where $P(x_1)$ is the distribution of the source symbols, $P(y_2, y_3 | x_1, x_3)$ the conditional channel distribution, i.e., the distribution of the channel outputs given the channel inputs and $P(\widehat{y}_3 | y_3, x_3)$ the "quantization rule" the relay has to choose. For a given s, the relay labels then $\widehat{y}_3^n(q, r | s)$ by

- $q \in [1, 2^{nR_3'}]$ (index within bin)

- $r \in [1, 2^{nR_3}]$ (bin index)

see Fig.3.7. The auxiliary random variable $\widehat{Y}_3$ represents a quantized version of Y_3 and the joint distribution of the random variables $X_1, X_3, Y_2, Y_3, \widehat{Y}_3$ factors as

$$P(x_1, x_3, y_2, y_3, \widehat{y}_3) = P(x_1) P(x_3) P(y_2, y_3 | x_1, x_3) P(\widehat{y}_3 | x_3, y_3). \tag{3.46}$$

Encoding at the Source. At the source terminal T_1 the message w of nRK bits is divided into K equally-sized blocks $w_1, w_2, \ldots, w_K$ of nR bits each. In block k, $k = 1, 2, \ldots, K+1$ terminal T_1 transmits $x_1^n(w_k)$ where $w_{K+1} = 1$.

Encoding at the Relay. In block $k = 1$, the relay terminal T_3 transmits $x_3^n(1)$, i.e., the first codeword from its codebook[6]. Right after block $k = 1$ (and just before block $k = 2$ starts) the

---

[6]Since in the first block the relay has not received anything from the source that could be helpful to decide which

relay observes y_{31}^n and tries to find a (q_1, r_1) from the quantization code book that is indexed by $s_1 = 1$, such that

$$(\widehat{y}_3^n(q_1, r_1 | s_1 = 1), y_{31}^n, x_3^n(s_1 = 1)) \in T_\epsilon^n\left(P_{\widehat{Y}_3 Y_3 X_3}\right), \tag{3.47}$$

where $T_\epsilon^n\left(P_{\widehat{Y}_3 Y_3 X_3}\right)$ denotes the jointly typical set with respect to $P_{\widehat{Y}_3 Y_3 X_3}$. The relay then sets $s_2 = r_1$, i.e., the bin index found in block $k = 1$ determines the codeword $x_3^n(s_2 = r_1)$ to be transmitted in block $k = 2$. Discussion of (3.47): The relay looks for a representation (quantization) $\widehat{y}_3^n$ of y_3^n such that both are jointly typical and both are jointly typical with the transmitted codeword x_3^n. Joint typicality between $\widehat{y}_3^n$ and y_3^n is necessary in order for $\widehat{y}_3^n$ being a good representation (quantization) of y_3^{n}[7]. Joint typicality between $\widehat{y}_3^n$ and x_3^n is necessary because the destination has to recover $\widehat{y}_3^n$ out of the transmitted x_3^n. Note the difference to rate distortion theory: there the decoder has direct access to the bin index of the quantized codeword $\widehat{y}_3^n$, here the index has to be encoded using the channel code x_3^n, such that the destination decoder can recover reliably the bin index of the quantized codeword.

After block $k = 2$ the relay observes y_{32}^n and tries to find a (q_2, r_2) from the quantization code book that is indexed by $s_2 = r_1$ such that

$$(\widehat{y}_3^n(q_2, r_2 | s_2 = r_1), y_{32}^n, x_3^n(s_2 = r_1)) \in T_\epsilon^n\left(P_{\widehat{Y}_3 Y_3 X_3}\right), \tag{3.48}$$

and sets $s_3 = r_2$. In general, after block k the relay observes y_{3k}^n and tries to find a (q_k, r_k) from the quantization code book that is indexed by $s_k = r_{k-1}$ such that

$$(\widehat{y}_3^n(q_k, r_k | s_k = r_{k-1}), y_{3k}^n, x_3^n(s_k = r_{k-1})) \in T_\epsilon^n\left(P_{\widehat{Y}_3 Y_3 X_3}\right). \tag{3.49}$$

We know from rate distortion theory that the relay is able to find a (q_k, r_k) with high probability when

$$R_3 + R_3' > I(\widehat{Y}_3; Y_3 | X_3) \tag{3.50}$$

and the block size n is large. The conditioning on X_3 in (3.50) is because the quantization code book from which we choose the $\widehat{y}_3^n$ is dependent on X_3. Roughly spoken, the relation (3.50) says that the relay should not compress y_3^n too much, otherwise the relay may not be able to find an appropriate quantization codeword that satisfies (3.49).

Decoding at the Destination. The destination starts after the completion of block $k = 2$.

---

codeword has to be sent to the destination.

[7]Recall rate distortion theory discussed in the previous chapter.

Remember, the relay has sent in this block the codeword $x_3^n(s_2 = r_1)$ that encodes the bin index r_1. This index gives information, in which bin the decoder can find $\hat{y}_3^n$ (which was chosen by the relay as quantization of y_{31}^n in block $k = 1$). The destination decoder looks for a $\tilde{s}_2$ such that

$$(x_3^n(\tilde{s}_2), y_{22}^n) \in T_\epsilon^n (P_{X_3 Y_2}) \tag{3.51}$$

and if successful puts out $r_1 = \tilde{s}_2$ as a result. The destination can decode r_1 as long as

$$R_3 < I(X_3; Y_2) \tag{3.52}$$

and n is large, i.e., the mutual information between relay and destination (treating the signal coming from the source as interference) has to be larger than the rate R_3 used by the relay to encode the bin index. Now the destination knows that $\hat{y}_3^n$ was chosen from code book $s_1 = 1$ (because in the first block $k = 1$, the relay has chosen $s_1 = 1$ per default) and that it lies in the bin addressed by the number r_1. However, the destination still does not know, which of the codewords in bin r_1 actually is the quantized version of y_3^n, i.e., it does not know the value of q_1. In order to resolve this uncertainty, the destination decoder considers the received symbols y_{21}^n in block $k = 1$ and looks for a $\tilde{q}_1$ such that

$$(\hat{y}_3^n(\tilde{q}_1, r_1 | s_1 = 1), y_{21}^n, x_3^n(s_1 = 1)) \in T_\epsilon^n (P_{\hat{Y}_3 Y_2 X_3}) \tag{3.53}$$

and if successful puts out $q_1 = \tilde{q}_1$ as a result. The destination can decode q_1 as long as

$$R_3' < I(\hat{Y}_3; Y_2 | X_3) \tag{3.54}$$

and n is large, i.e., the number of codewords in one bin should be less than the mutual information between the receive signal y_2^n and the signal $\hat{y}_3^n$ to be decoded. The conditioning on X_3 is again because the statistics of a quantization code book used in block k depends on the codeword x_3^n that is sent in block k.

In general, after block k the destination terminal T_2 tries to estimate the bin index r_{k-1} by looking for a $\tilde{s}_k$ such that

$$(x_3^n(\tilde{s}_k), y_{2k}^n) \in T_\epsilon^n (P_{X_3 Y_2}) \tag{3.55}$$

and if successful puts out $r_{k-1} = \tilde{s}_k$. Terminal T_2 succeeds with high probability if

$$R_3 < I(X_3; Y_2) \tag{3.56}$$

61

and the block size n is large. Relation (3.56) says, that the bin index $s_k = r_{k-1}$ (*coarse* information about the quantized relay observation) is being communicated reliably to terminal T_2, i.e., the destination terminal knows r_{k-1} after block k. Terminal T_2 then uses $y_{2(k-1)}^n$ to find a q_{k-1} (*fine* information about the quantized relay observation) such that

$$(\widehat{y}_3^n(\widetilde{q}_{k-1}, r_{k-1}|s_{k-1} = r_{k-2}), y_{2(k-1)}^n, x_3^n(s_{k-1} = r_{k-2})) \in T_\epsilon^n\left(P_{\widehat{Y}_3 Y_2 X_3}\right) \qquad (3.57)$$

and if successful[8] puts out $q_{k-1} = \widetilde{q}_{k-1}$. Terminal T_2 succeeds with high probability when

$$R_3' < I(\widehat{Y}_3; Y_2|X_3) \qquad (3.58)$$

and the block size n is large. This means that there shouldn't be too much codewords in one bin, otherwise the decoder at terminal T_2 is not able to resolve the uncertainty (to decode the *fine* information q) with help of the side information Y_2.

Finally, terminal T_2 uses both $y_{2(k-1)}^n$ and $\widehat{y}_3^n(q_{k-1}, r_{k-1}|s_{k-1})$ to find an index $\widetilde{w}_{k-1}$ such that

$$(x_1^n(\widetilde{w}_{k-1}), \widehat{y}_3(q_{k-1}, r_{k-1}|s_{k-1} = r_{k-2}), y_{2(k-1)}^n, x_3^n(s_{k-1} = r_{k-2})) \in T_\epsilon^n\left(P_{X_1 \widehat{Y}_3 Y_2 X_3}\right) \quad (3.59)$$

and if successful puts out $w_{k-1} = \widetilde{w}_{k-1}$. Terminal T_2 succeeds with high probability when

$$R < I(X_1; \widehat{Y}_3 Y_2|X_3) \qquad (3.60)$$

and the block size n is large. By choosing

$$R_3' = I(\widehat{Y}_3; Y_2|X_3) - \epsilon \qquad (3.61)$$

we get from (3.50)

$$R_3 > I(\widehat{Y}_3; Y_3|X_3) - I(\widehat{Y}_3; Y_2|X_3) + \epsilon \qquad (3.62)$$
$$= H(\widehat{Y}_3|X_3) - H(\widehat{Y}_3|X_3 Y_3) - H(\widehat{Y}_3|X_3) + H(\widehat{Y}_3|X_3 Y_2) + \epsilon \qquad (3.63)$$
$$= H(\widehat{Y}_3|X_3 Y_2) - H(\widehat{Y}_3|X_3 Y_2 Y_3) + \epsilon \qquad (3.64)$$
$$= I(\widehat{Y}_3; Y_3|X_3 Y_2) + \epsilon \qquad (3.65)$$

where in the first step we expressed the mutual information expressions in terms of entropies,

---

[8]One needs the Markov Lemma 2.1.5 in order to show that the probability of (3.59) goes to one as n grows large. We will provide the proof in the next chapter, where we extend the CF scheme to two-way relaying.

and in the second step we used the fact that $H(\widehat{Y}_3|X_2Y_3) = H(\widehat{Y}_3|X_2Y_2Y_3)$, since $\widehat{Y}_3|Y_3$ does not change its statistics when Y_2 is known at the decoder. The achievable rate of the compress-and-forward strategy as stated in Theorem 3.40 is then given in (3.60), and the constraint (3.41) follows by combining (3.56) and (3.65). □

Next we evaluate the achievable rate in Proposition 3.1.5 for the case of an AWGN relay channel. The probability distribution that maximizes (3.40) is not known. We therefore choose

$$p(x_1) \sim \mathcal{CN}(0, P_1) \tag{3.66}$$

$$p(x_3) \sim \mathcal{CN}(0, P_3) \tag{3.67}$$

with X_1, X_3 statistically independent. The auxiliary random variable $\widehat{Y}_3$ is defined as

$$\widehat{Y}_3 = Y_3 + Z_c = X_1 h_1 + Z_3 + Z_c \tag{3.68}$$

with $Z_c \sim \mathcal{CN}(0, \sigma_c^2)$ and independent of X_1 and X_3, i.e., σ_c^2 is the variance of the compression (quantization) noise. The achievable rate follows then as

$$I(X_1; Y_2\widehat{Y}_3|X_3) = \log\left(1 + \frac{P_1|h_0|^2}{\sigma_2^2} + \frac{P_1|h_1|^2}{\sigma_3^2 + \sigma_c^2}\right), \tag{3.69}$$

i.e., it corresponds to a point-to-point channel with one transmit antenna and two receive antennas where at the second receive antenna there is some additional noise Z_c (assuming $\sigma_2^2 = \sigma_3^2$). The reason is that the observation at the "second receive antenna" is not known perfectly, but within some distortion σ_c^2. It remains to determine the variance of the compression noise σ_c^2. From (3.62) we have

$$R_3 > I(\widehat{Y}_3; Y_3|X_3) - I(\widehat{Y}_3; Y_2|X_3) + \epsilon \tag{3.70}$$

$$= h(\widehat{Y}_3|X_3Y_2) - h(\widehat{Y}_3|X_3Y_3) + \epsilon \tag{3.71}$$

$$= h(Y_3 + Z_c, Y_2|X_3) - h(Y_2|X_3) - h(Z_c) + \epsilon \tag{3.72}$$

$$= \log\left(\pi e|R_{\widehat{Y}_3,Y_2}|\right) - h(Y_2|X_3) - h(Z_c) + \epsilon \tag{3.73}$$

where the covariance matrix $R_{\widehat{Y}_3,Y_2}$ is given as

$$R_{\widehat{Y}_3,Y_2} = \begin{pmatrix} \sigma_{\widehat{Y}_3}^2 + \sigma_c^2 & \rho \\ \rho^* & \sigma_{Y_2}^2 \end{pmatrix} \tag{3.74}$$

with

$$\rho = \mathcal{E}\left\{(Y_3 + Z_c)Y_2^*\right\} = Ph_0h_1^*. \tag{3.75}$$

It follows

$$R_3 > \log\left((P_1|h_1|^2 + \sigma_3^2 + \sigma_c^2)(P_1|h_0|^2 + \sigma_2^2) - P_1^2|h_0|^2|h_1|^2\right) \tag{3.76}$$

$$- \log\left(P_1|h_0|^2 + \sigma_2^2\right) - \log\left(\sigma_c^2\right) + \epsilon \tag{3.77}$$

and from that we get

$$\sigma_c^2 > \frac{P_1|h_0|^2\sigma_3^2 + P_1|h_1|^2\sigma_2^2 + \sigma_2^2\sigma_3^2}{(2^{R_3} - 1)\left(P_1|h_0|^2 + \sigma_2^2\right)}. \tag{3.78}$$

The rate R_3 follows by the constraint (3.41) which says that the destination has to decode the bin index of the chosen quantization codeword $\hat{y}_3^n(q, r|s)$ without errors. Hence,

$$R_3 \leq I(X_3; Y_2) = \log\left(1 + \frac{P_3|h_2|^2}{\sigma_2^2 + P_1|h_0|^2}\right), \tag{3.79}$$

i.e., the destination decodes the bin index by treating the signal from the source as interference. We summarize the results in the following proposition:

Proposition 3.1.6 (Achievable Rate for AWGN Compress-and-forward). *An achievable rate for the AWGN relay channel is given by*

$$R \leq \log\left(1 + \frac{P_1|h_0|^2}{\sigma_2^2} + \frac{P_1|h_1|^2}{\sigma_3^2 + \sigma_c^2}\right) \tag{3.80}$$

where

$$\sigma_c^2 > \frac{P_1|h_0|^2\sigma_3^2 + P_1|h_1|^2\sigma_2^2 + \sigma_2^2\sigma_3^2}{(2^{R_3} - 1)\left(P_1|h_0|^2 + \sigma_2^2\right)} \tag{3.81}$$

and

$$R_3 \leq \log\left(1 + \frac{P_3|h_2|^2}{\sigma_2^2 + P_1|h_0|^2}\right). \tag{3.82}$$

In Section 4.6 we will compare the decode-and-forward scheme with the compress-and-forward scheme for a simple relay network geometry. But first, we evaluate in the next section the achievable rate for an amplify-and-forward strategy.

3.1.5 Amplify-and-forward

This strategy is very popular since it requires no sophisticated signal processing at the relay. The relay just amplifies the received signal according to its transmit power constraint and retransmits the signal to the destination. We analyze the achievable rate directly for an AWGN relay channel. The receive signal at the destination terminal in channel use k is given as

$$Y_2[k] = h_0 X_1[k] + g h_1 h_2 X_1[k-1] + g h_2 Z_3[k-1] + Z_2[k] \qquad (3.83)$$

where the relay gain is chosen as

$$g = \sqrt{\frac{P_3}{P_1 |h_1|^2 + \sigma_3^3}}, \qquad (3.84)$$

i.e., the relay scales $Y_3[k]$ such that the transmit power is P_3. We see from (3.83) that the relay channel with an amplify-and-forward relay terminal turns the equivalent channel into a one-tap inter-symbol interference (ISI) channel with impulse response

$$h[k] = h_0 \delta[k] + g h_1 h_2 \delta[k-1] \qquad (3.85)$$

where the compound noise at the destination

$$\widetilde{Z}_2[k] = g h_2 Z_3[k-1] + Z_2[k] \qquad (3.86)$$

has the autocorrelation function

$$\mathbb{E}\left[\widetilde{Z}_2[k]\widetilde{Z}_2^*[m]\right] = \left(\sigma_2^2 + \sigma_3^3 g^2 |h_2|^2\right)\delta[k-m]. \qquad (3.87)$$

The achievable rate (3.83) is determined in the frequency-domain[9]:

$$
\begin{aligned}
R &= \frac{1}{2\pi}\int_{-\pi}^{\pi} \log\left(1 + \frac{P_1 |H(\omega)|^2}{\sigma_2^2 + \sigma_3^3 g^2 |h_2|^2}\right) d\omega \\
&= \frac{1}{2\pi}\int_{-\pi}^{\pi} \log\left(1 + \frac{P_1}{\sigma_2^2 + \sigma_3^3 g^2 |h_2|^2}\left(\left|h_0 + g h_1 h_2 e^{-j\omega}\right|^2\right)\right) d\omega \\
&= \log\left(\frac{1 + a + \sqrt{(1+a)^2 - b^2}}{2}\right) \qquad (3.88)
\end{aligned}
$$

---

[9]For simplicity we choose a symbol interval $T = 1$.

where

$$a = \frac{P_1 \left(|h_0|^2 + g^2|h_1|^2|h_2|^2 \right)}{\sigma_2^2 + \sigma_3^3 g^2 |h_2|^2},$$

(3.89)

$$b = \frac{2P_1 g |h_1||h_2|}{\sigma_2^2 + \sigma_3^3 g^2 |h_2|^2}.$$

(3.90)

Note that it may be difficult to implement a complete analog relay, i.e., an analog circuit that maps an $s(t) + n(t)$ to $g \cdot (s(t) + n(t))$. In practice one could use an analog-to-digital converter to map the analog signal into the digital domain and to perform the amplification digitally and then to use a digital-to-analog converter to re-transform the signal into the analog domain.

In principal the amplify-and-forward strategy needs no digital signal processing whereas decode-and-forward as well as compress-and-forward need sophisticated signal processing algorithms at the relay. Compared to these strategies a reasonable assumption for amplify-and-forward could be to neglect the delay incurred at the AF relay. The receive signal at the destination is then

$$Y_2[k] = h_0 X_1[k] + g h_1 h_2 X_1[k] + g h_2 Z_3[k] + Z_2[k],$$

(3.91)

i.e., the signal contributions of the source $h_0 X_1[k]$ and the relay $g h_1 h_2 X_1[k]$ arrive at the same time at the destination. In comparison to (3.83) there is no intersymbol interference anymore. The achievable rate is simply

$$R(g) = \log\left(1 + \frac{P_1 |h_0 + g h_1 h_2|^2}{\sigma_2^2 + \sigma_3^2 |g h_2|^2} \right).$$

(3.92)

We can choose g such that the rate $R(g)$ is maximized,

$$g_{\text{opt}} = \sqrt{\frac{P_3}{P_1 |h_1|^2 + \sigma_3^2}} \left| \frac{h_0^* h_2}{h_1^*} \right| \frac{h_1^*}{h_0^* h_2}$$

(3.93)

which follows by setting the derivative of (3.92) with respect to the complex g to zero and scaling the optimal g such, that the transmit power of the relay is P_3.

3.1.6 Numerical Examples

We compare now the achievable rates obtained for decode-and-forward, compress-and-forward and amplify-and-forward for a simplified relay network as depicted in Fig.3.8. The channel

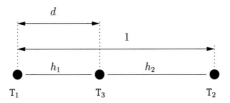

Fig. 3.8: Simplified relay network

between the source terminal and destination terminal is normalized to $h_0 = 1$. The channel between source and relay is chosen as $h_1 = 1/d^{\delta/2}$ and the channel between relay and destination as $h_2 = 1/(1 - d)^{\delta/2}$. The parameter δ denotes the path loss exponent and is chosen to be 2 (free space propagation). We assume $P_1 = P_3 = 10$ and $\sigma_2^2 = \sigma_3^2 = 1$. In Fig.3.9 we see the achievable rates of the different strategies compared to the cut-set upper bound (3.39) and the case where the relay is switched off, where the source-destination capacity is

$$C_{\text{relay}-\text{off}} = \log\left(1 + \frac{P|h_0|^2}{\sigma_2^2}\right). \tag{3.94}$$

Note that in this case the relay has still transmit power P available, i.e., the the two-node "network" consumes only power P in each time slot. When the relay is active the three-node network power consumption is $2P$ in each time slot, i.e, we compare on the basis of a node power constraint rather than on a network power constraint.

We observe from Fig.3.9 that the DF strategy achieves the cut–set upper bound when the relay is in the vicinity of the source, i.e., the source-relay channel is strong. The reason is, that in this case the multiple access capacity (source and relay to destination) is the bottleneck for the capacity of the relay channel. The multiple access capacity dominates then the cut-set bound as well as the DF lower bound. When the relay is in the vicinity of the destination terminal, the CF strategy achieves the cut-set upper bound. In this case the broadcast capacity (source to relay and destination) is the bottleneck for the capacity of the relay channel and this capacity dominates then the cut–set bound as well as the CF lower bound. We further observe, that the CF strategy always outperforms the direct transmission capacity (this is not true, when both strategies operate with the same network power, i.e, when in the relay-off case the source sends with power $2P$), whereas the DF strategy is worse than direct transmission, when the source-relay capacity is

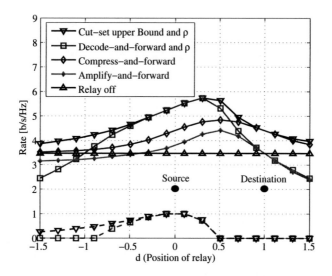

Fig. 3.9: Achievable rates for different relaying strategies and corresponding correlation coefficients for the cut-set bound and the DF strategy

below the source-destination capacity, i.e.,

$$\log\left(1 + \frac{P|h_1|^2}{\sigma_3^3}\right) < \log\left(1 + \frac{P|h_0|^2}{\sigma_2^2}\right). \tag{3.95}$$

We further note that the AF and CF strategy achieve their best performance when the relay is half-way between source and destination. For the case that the relay is near by the destination, the DF and the AF strategy converge to the same performance as can be seen by letting $h_2 \to \infty$ in (3.33) and (3.88).

In Fig.3.10 we compare the amplify-and-forward protocols with and without delay at the relay. We see that the relay without delay (with optimized g) significantly outperforms the case where the relay is one time slot behind. Note that for some relay positions the rate of the relay without delay is even larger than the cut-set bound. How is this possible? The reason is, that the cut-set bound is based on the assumption, that the relay can only use past received symbols, but not the current symbol in order to determine its transmit signal. In [100] the authors have generalized the cut-set bound to include the case where the relay can use also the current received symbol. For AWGN channels, the generalized cut-set bound is

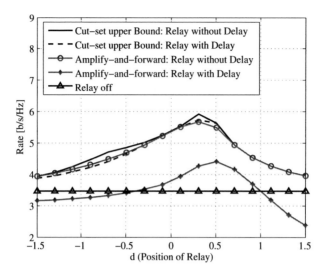

Fig. 3.10: Achievable rates for amplify-and-forward with and without delay at the relay

Proposition 3.1.7 (Cut-set Upper Bound for AWGN Relay Channel without Delay).

$$C \leq \max_{0 \leq \rho_r \leq 1} \min \left(C \left(P_1 \left(\frac{|h_0|^2}{\sigma_2^2} + \frac{|h_1|^2}{\sigma_3^2} - \frac{\rho_r^2}{1 + \frac{P_1}{\sigma_3^2}|h_1|^2(1-\rho_r^2)} \right) \right), \right.$$
$$\left. C \left(\frac{P_1|h_0|^2 + P_3|h_2|^2 + 2\rho_r\sqrt{P_1 P_3}|h_0||h_2|}{\sigma_2^2} \right) \right). \tag{3.96}$$

A proof can be found in [100]. Note that the multiple access rate (the second term in (3.96) did not change compared to the classical cut-set bound (3.39), only the broadcast rate is different when we allow the relay to use the current symbol too in order to determine it's transmit signal. In Fig.3.10 we see that the generalized cut-set bound is slightly larger than the ordinary cut-set bound, and we also observe, that the amplify-and-forward relay strategy with no delay is always below the general cut-set bound (3.96). Nevertheless, the AF strategy seems to perform extremely well since it is close to the cut-set bound at all relay positions.

3.2 Application to Fading Relay Channels

In this section we present the extension of the relaying strategies discussed in the previous section to fading relay channels. Section 3.2.2 treats the case of full-duplex fading relay channels, i.e., we still assume that the relay can receive and transmit at the same time in the same frequency band. This may be achieved for example with two antennas, one for reception and one for transmission. In order to prevent strong self-interference the antennas have to be decoupled sufficiently. Section 3.2.3 presents capacity theorems for the more practical half-duplex fading relay channel. A half-duplex relay cannot receive and transmit simultaneously in the same frequency channel. For each time slot T the relay receives during αT and then transmits during $(1 - \alpha)T$. In principle the separation could also be done in the frequency domain. However, as pointed out in [16] time division duplexing has the advantage that α may be adapted with respect to instantaneous channel knowledge whereas fractional bandwidths assigned to the transmit and receives modes of the relay are usually fixed. The reason is that adjusting the filter bandwidth is practically more involved than adjusting the switching time between receiving and transmitting.

3.2.1 System Model

Consider again the relay channel depicted in Fig.3.6. We now assume that the channel coefficients h_0, h_1 and h_2 are complex random variables and change i.i.d. from coherence interval to coherence interval. We rewrite the receive signals at the destination and relay as[10]

$$y_2 = x_1 c_0 e^{j\phi_0} + x_3 c_2 e^{j\phi_2} + z_2 \tag{3.97}$$

$$y_3 = x_1 c_1 e^{j\phi_1} + z_3, \tag{3.98}$$

where $h_i = c_i e^{j\phi_i}$ for $i = 0, 1, 2$. We assume that a transmitted codeword captures enough coherence intervals in order to average out the fading and that each coherence interval is long enough in order to average out thermal noise disturbances. Further we assume that all nodes are symbol synchronous but not necessarily phase synchronous. Carrier synchronization between source and relay requires oscillators at both nodes that are phase-locked and this might be a very challenging task in practice. For the asynchronous phase model one can incorporate a random phase offset into the source-relay channel phase.

In the following we consider the ergodic capacities of the strategies discussed in Section 3.1

---

[10]When dealing with fading channels we denote random variables as well as realizations of random variables by lower-case letters.

for the case of full-duplex relaying as well as half-duplex relaying. We assume that the source and the relay do not have transmitter channel state information (CSIT) but know the channel distributions. We assume that all nodes have full receiver channel state information (CSIR).

Another performance measure in fading channels is outage capacity, where it is assumed that the fading coefficients are fixed during the transmission of a codeword. Outage occurs when the chosen transmission rate is above the current mutual information of the channel. Outage capacity or outage probability is an important measure when we compare different transmission schemes with respect to the achievable diversity gain. Since in this work we are primarily interested in achievable rates of cooperative systems we concentrate on ergodic capacity. The outage performance of decode-, compress- and amplify-and-forward relay channels is investigated in [102].

3.2.2 Wireless Full-duplex Relaying

Remember that the relay has full knowledge of $h_1 = c_1 e^{j\phi_1}$ and the destination has full knowledge of $h_0 = c_0 e^{j\phi_0}$ and $h_2 = c_2 e^{j\phi_2}$. We further assume that the magnitudes and phases are independent and that the phases are uniformly distributed over $[0, 2\pi)$. The cut-set upper bound follows as

$$C \leq \max_{\rho \in \mathbb{C}} \min \left(C_1, C_2 \right) \tag{3.99}$$

where

$$C_1 = \mathcal{E} \left\{ \log \left(1 + P_1 (1 - |\rho|^2) \left(\frac{c_0^2}{\sigma_2^2} + \frac{c_1^2}{\sigma_3^2} \right) \right) \right\} \tag{3.100}$$

$$C_2 = \mathcal{E} \left\{ \log \left(1 + \frac{P_1 c_0^2 + P_3 c_2^2 + 2\mathfrak{Re} \left\{ \rho c_0 c_2 e^{j(\phi_0 - \phi_2)} \right\} \sqrt{P_1 P_3}}{\sigma_2^2} \right) \right\}. \tag{3.101}$$

We first look at the correlation coefficient $\rho \in \mathbb{C}$ that maximizes C_2, i.e.,

$$C_2^* = \max_{\rho \in \mathbb{C}} \mathcal{E} \left\{ \log \left(1 + \frac{P_1 c_0^2 + P_3 c_2^2 + 2|\rho| \cos \left(\phi_0 - \phi_2 + \arg(\rho) \right) \sqrt{P_1 P_3}}{\sigma_2^2} \right) \right\}$$

$$\leq \max_{\rho \in \mathbb{C}} \log \left(1 + \mathcal{E} \left\{ \frac{P_1 c_0^2 + P_3 c_2^2 + 2|\rho| \cos \left(\phi_0 - \phi_2 + \arg(\rho) \right) \sqrt{P_1 P_3}}{\sigma_2^2} \right\} \right)$$

$$= \log \left(1 + \frac{P_1 \mathcal{E} \left\{ c_0^2 \right\} + P_3 \mathcal{E} \left\{ c_2^2 \right\}}{\sigma_2^2} \right) \tag{3.102}$$

where in the second inequality we used Jensen's inequality [74]. From (3.102) it follows that $\rho = 0$ maximizes C_2. Since C_1 can only be decreased for $|\rho| \neq 0$ it follows that $\rho = 0$ optimizes the cut-set bound for the ergodic capacity.

Proposition 3.2.1 (Cut-set Upper Bound for Ergodic Fading Relay Channel)**.**

$$C \leq \min \left(\mathcal{E} \left\{ C \left(P_1 \left(\frac{c_0^2}{\sigma_2^2} + \frac{c_1^2}{\sigma_3^2} \right) \right) \right\}, \mathcal{E} \left\{ C \left(\frac{P_1 c_0^2 + P_3 c_2^2}{\sigma_2^2} \right) \right\} \right) \qquad (3.103)$$

The intuition behind the optimality of $\rho = 0$ is that since the source and the relay do not have CSIT they cannot choose ρ such that the transmitted signals from the source and the relay add up coherently at the destination. Using the same arguments we also obtain the ergodic rate of the decode-and-forward scheme:

Proposition 3.2.2 (Ergodic Rate for Decode-and-forward)**.**

$$R = \min \left(\mathcal{E} \left\{ C \left(\frac{P_1 c_1^2}{\sigma_3^2} \right) \right\}, \mathcal{E} \left\{ C \left(\frac{P_1 c_0^2 + P_3 c_2^2}{\sigma_2^2} \right) \right\} \right) \qquad (3.104)$$

From (3.103) and (3.104) we observe that the decode-and-forward strategy achieves the cut-set upper bound and therefore the ergodic capacity of the fading relay channel when [15]

$$\mathcal{E} \left\{ C \left(\frac{P_1 c_0^2 + P_3 c_2^2}{\sigma_2^2} \right) \right\} < \mathcal{E} \left\{ C \left(\frac{P_1 c_1^2}{\sigma_3^2} \right) \right\}. \qquad (3.105)$$

The compress-and-forward scheme is not directly applicable to fading channels when the relay has only receiver CSI. Since the side information at the destination depends on the source-destination channel gain h_0 the relay has to know this channel gain in order to choose the variance of the compression noise σ_c^2 (3.81) appropriately. The outage capacity of the compress-and-forward scheme for both, known and unknown h_0 at the relay, has been derived in [102]. For ergodic fading channels with full CSI the compress-and-forward strategy follows directly from the AWGN case. From (3.77) we choose

$$R_3 = \mathcal{E} \left\{ \log \left((P_1 c_1^2 + \sigma_3^2 + \sigma_c^2)(P_1 c_0^2 + \sigma_2^2) - P_1^2 c_0^2 c_1^2 \right) \right\} \qquad (3.106)$$

$$- \mathcal{E} \left\{ \log \left(P_1 c_0^2 + \sigma_2^2 \right) - \log \left(\sigma_c^2 \right) \right\}. \qquad (3.107)$$

At the same time the rate R_3 has to be smaller or equal than the capacity of the relay-destination channel, which gives us

$$C_{\text{rd}} = \mathcal{E} \left\{ \log \left(1 + \frac{P_3 c_2^2}{\sigma_2^2 + P_1 c_0^2} \right) \right\}. \qquad (3.108)$$

Equalizing (3.107) and (3.108) gives us an equation for the variance of the compression noise

σ_c^2. For Rayleigh fading and a high-SNR approximation σ_c^2 was determined in [16, Appendix E]. In general, the variance depends on the fading distribution of the channel coefficients. Another solution for the ergodic compress-and-forward strategy was presented in [84], where the CF scheme does not take advantage of the side information at the destination, i.e., quantization at the relay is done without taking the side information into account. The ergodic rate of the amplify-and-forward scheme is obtained by simply taking the expectation of (3.88) with respect to all channel coefficients. Numerical examples are provided in Section 3.2.4.

3.2.3 Wireless Half-duplex Relaying

The previous discussions considered a full-duplex relay, i.e., the relay was able to receive and transmit at the same time and frequency. Such a relay may be implemented, when two antennas are used for transmission and reception. However, since incoming signals are very weak and outgoing signals very strong[11] the transmitted signal will strongly interfere with the received signal. Canceling the self-interference may be extremely difficult when there is no isolation between the two antennas, since the analog amplifiers in the receive chain may be driven into saturation and cause very strong nonlinear distortions to the received signal. A more feasible solution and easier to implement would be to use a half-duplex relay that transmits and receives on different channels that are separated in frequency (frequency division, FD) or time (time division, TD). In the TD mode the relay receives during the time αT and transmits during the time $(1 - \alpha)T$. In the FD mode the relay receives in the bandwidth αW and transmits in the bandwidth $(1 - \alpha)W$. As pointed out in [16] the TD mode and FD mode are equivalent from an information-theoretic point of view. However, the TD mode has the advantage, that it is easier to adjust α to instantaneous channel conditions in the TD mode than in the FD mode. We therefore consider only the TD mode.

In this section the results from the previous section are applied to half-duplex TD relaying. The cut-set upper bound, decode-and-forward and compress-and-forward strategies were obtained in [16]. The amplify-and-forward strategy was investigated in [63], [41], [38].

Assume that the source node transmits with power $P_1^{(1)}$ in the relay-receive period (first time slot) and with power $P_1^{(2)}$ in the relay-transmit period (second time slot). The relay transmits with power P_3. The following is an upper bound on the capacity of the half-duplex TD AWGN relay channel with deterministic channel coefficients h_0, h_1 and h_2.

---

[11]80-100 dB difference in signal power is typical in practice.

Proposition 3.2.3 (Cut-set Upper Bound for Half-duplex AWGN Relay Channel).

$$C \le \max_{0 \le \alpha, \beta \le 1} \min \left(C_1(\alpha, \beta), C_2(\alpha, \beta) \right) \tag{3.109}$$

$$C_1(\alpha, \beta) = \alpha \log \left(1 + P_1^{(1)} \left(c_0^2 + c_1^2 \right) \right)$$
$$+ (1 - \alpha) \log \left(1 + (1 - \beta) P_1^{(2)} c_0^2 \right) \tag{3.110}$$

$$C_2(\alpha, \beta) = \alpha \log \left(1 + P_1^{(2)} c_0^2 \right) + (1 - \alpha) \log \left(1 + P_1^{(2)} c_0^2 \right.$$
$$\left. + P_3 c_2^2 + 2\beta \sqrt{P_1^{(2)} P_3 c_0 c_2} \right) \tag{3.111}$$

where β is the real part of the correlation coefficient ρ defined in (3.21). The proof follows by applying the cut-set bound [74, Theorem 14.10.1]. The details are given in [16, Appendix A]. For the case of ergodic fading coefficients $h_i = c_i e^{j\phi_i}$, $i = 0, 1, 2$ and no CSI at transmitters (besides knowledge of the fading statistics) the upper bound is maximized for $\beta = 0$, because source and relay cannot co-phase their transmit signals.

Proposition 3.2.4 (Cut-set Upper Bound for Half-duplex Fading Relay Channel).

$$C \le \max_{0 \le \alpha \le 1} \min \left(C_1(\alpha), C_2(\alpha) \right) \tag{3.112}$$

$$C_1(\alpha) = \mathcal{E} \left\{ \alpha \log \left(1 + P_1^{(1)} \left(c_0^2 + c_1^2 \right) \right) + (1 - \alpha) \log \left(1 + P_1^{(2)} c_0^2 \right) \right\} \tag{3.113}$$

$$C_2(\alpha) = \mathcal{E} \left\{ \alpha \log \left(1 + P_1^{(2)} c_0^2 \right) + (1 - \alpha) \log \left(1 + P_1^{(2)} c_0^2 + P_3 c_2^2 \right) \right\} \tag{3.114}$$

Note that the relay-transmit and relay-receive periods (adjusted by α) are determined based on channel statistics, not on instantaneous channel conditions. When the relay is enforced to decode the message sent by the source, following decode-and-forward rate is achievable for AWGN channels:

Proposition 3.2.5 (Achievable Rate for Decode-and-forward Half-duplex AWGN Relay Channel).

$$R = \max_{0 \le \alpha, \beta \le 1} \min \left(R_1(\alpha, \beta), R_2(\alpha, \beta) \right) \tag{3.115}$$

$$R_1(\alpha, \beta) = \frac{\alpha}{2} \log \left(1 + P_1^{(1)} c_1^2 \right)$$
$$+ \frac{1 - \alpha}{2} \log \left(1 + (1 - \beta) P_1^{(2)} c_0^2 \right) \tag{3.116}$$

$$R_2(\alpha, \beta) = \frac{\alpha}{2} \log \left(1 + P_1^{(2)} c_0^2 \right) + \frac{1 - \alpha}{2} \log \left(1 + P_1^{(2)} c_0^2 \right.$$
$$\left. + P_3 c_2^2 + 2\beta \sqrt{P_1^{(2)} P_3 c_0 c_2} \right) \tag{3.117}$$

For the achievability of Proposition (3.2.5) the block Markov structure in the encoding process is not needed as in the full-duplex case: the message w at the source is split into two parts, w_r and w_d. During the relay-receive period αT the source transmits w_r which is observed by the relay and the destination. During the relay-transmit period $(1 - \alpha)T$ the source transmits a superposition of w_r and w_d and the relay transmits w_r (which was previously decoded correctly by the relay). The destination decodes first w_r using the observations from both time slots, hereby treating w_d as interference. Hereby, the signals from the first and the second time slot add up coherently. The destination then subtracts the influence of w_r and decodes w_d using the observation from the second time slot. Hence, w_d is transmitted without the help of the relay and w_r is transmitted with the help of the relay. The full proof is given in [16, Appendix A]. The achievable rate for ergodic fading channels follows in the same way as for the cut-set upper bound, i.e., $\beta = 0$ and taking the average over all channel realisations, as in Proposition 3.2.4.

The compress-and-forward strategy can also be extended to the half-duplex case:

Proposition 3.2.6 (Achievable Rate for Compress-and-forward Half-duplex AWGN Relay Channel).

$$R = \frac{\alpha}{2} \log \left(1 + P_1^{(1)} c_0^2 + \frac{P_1^{(1)} c_1^2}{1 + \sigma_c^2} \right) + \frac{1 - \alpha}{2} \log \left(1 + c_0^2 P_1^{(2)} \right) \tag{3.118}$$

where the compression noise is given by

$$\sigma_c^2 = \frac{P_1^{(1)} \left(c_1^2 + c_0^2 \right) + 1}{\left(\left(1 + \frac{P_3 c_2^2}{1 + P_1^{(2)} c_0^2} \right)^{(1-\alpha)/\alpha} - 1 \right) \left(P_1^{(1)} c_0^2 + 1 \right)} \tag{3.119}$$

The proof is given in [16]. Again, the rate for fading channels follows in the same way as for the full-duplex case [16].

For the half-duplex amplify-and-forward scheme different transmission protocols are possible [38, 41, 42, 63]:

Protocol I: The source transmits in the period αT and is quiet in the period $(1 - \alpha)T$, whereas the relay receives during αT, amplifies its received signal with a specific gain g and transmits in the period $(1 - \alpha)T$. The destination uses then its observations in both time slots to decode the data. This protocol was proposed in [63].

Protocol II: This protocol allows the source to be active in the second time slot and to transmit new information symbols during the relay-transmit period $(1 - \alpha)T$. This leads to a multiple access channel in the second time slot (relay and source communicate with the destination). This protocol was proposed in [41] and [38]. In [38] also a third protocol was discussed, where the destination does not use its observation in the first time slot and it was shown that the second protocol leads to the highest spectral efficiency of all three protocols. We discuss here the achievable rates of the first and the second protocol. We assume equal lengths of the relay-receive and relay-transmit periods, i.e., $\alpha = 1/2$. For Protocol I the input-output relation during the relay-receive period in time slot k is defined as

$$y_2[k] = x_1[k]h_0 + z_2[k] \tag{3.120}$$

$$y_3[k] = x_1[k]h_1 + z_3[k]. \tag{3.121}$$

During the relay-transmit period we have

$$y_2[k + 1] = x_1[k]h_1 g h_2 + g(\alpha)h_2 z_3[k] + z_2[k + 1] \tag{3.122}$$

$$y_3[k + 1] = 0 \tag{3.123}$$

where we assume that $\mathcal{E}\{|x_1|^2\} = P_1 = P$ and $\mathcal{E}\{|x_3|^2\} = P_3 = P$. We then choose for the relay gain

$$g = \sqrt{\frac{P_3}{P_1|h_1|^2 + \sigma_3^2}} = \sqrt{\frac{P}{P|h_1|^2 + \sigma_3^2}}. \tag{3.124}$$

Note that with these power normalizations each node consumes a transmit energy $PT/2$ in one time slot of length $T/2$. In vector notation we obtain [63]

$$\mathbf{y} = \begin{pmatrix} y_2[k] \\ y_2[k+1] \end{pmatrix} = \underbrace{\begin{pmatrix} h_0 \\ h_1 g h_2 \end{pmatrix}}_{\mathbf{H}} x_1[k] + \underbrace{\begin{pmatrix} 0 & 1 & 0 \\ g h_2 & 0 & 1 \end{pmatrix}}_{\mathbf{R}^{\frac{1}{2}}} \underbrace{\begin{pmatrix} z_3[k] \\ z_2[k] \\ z_2[k+1] \end{pmatrix}}_{\mathbf{z}}. \tag{3.125}$$

The achievable rate for Protocol I follows by applying the results from [9], [10]:

$$R_1 = \frac{1}{2} \log \det \left(\mathbf{I}_2 + P\mathbf{R}^{-1} \mathcal{E} \left\{ \mathbf{z}\mathbf{z}^{\mathrm{H}} \right\} \mathbf{H}\mathbf{H}^{\mathrm{H}} \right) \tag{3.126}$$

$$= \frac{1}{2} \log \left(1 + \frac{P|h_0|^2}{\sigma_2^2} + \frac{P|h_1 g h_2|^2}{(|g h_2|^2 \sigma_3^2 + \sigma_2^2)} \right). \tag{3.127}$$

We now turn our attention to Protocol II. The input-output relation during the relay-receive period in time slot k is defined as

$$y_2[k] = x_1[k]h_0 + z_2[k] \tag{3.128}$$

$$y_3[k] = x_1[k]h_1 + z_3[k]. \tag{3.129}$$

During the relay-transmit period we have

$$y_2[k+1] = x_1[k+1]h_0 + x_1[k]h_1 g h_2 + g h_2 z_3[k] + z_2[k+1] \tag{3.130}$$

$$y_3[k+1] = 0 \tag{3.131}$$

where we again assume that $\mathcal{E}\left\{|x_1|^2\right\} = P_1 = P$ and $\mathcal{E}\left\{|x_3|^2\right\} = P_3 = P$. Again the energy consumption for both source and relay is $PT/2$ per time slot. In vector notation we have [41], [38]

$$\mathbf{y} = \begin{pmatrix} y_2[k] \\ y_2[k+1] \end{pmatrix} = \underbrace{\begin{pmatrix} h_0 & 0 \\ h_1 g h_2 & h_0 \end{pmatrix}}_{\mathbf{H}} \begin{pmatrix} x_1[k] \\ x_1[k+1] \end{pmatrix}$$

$$+ \underbrace{\begin{pmatrix} 0 & 1 & 0 \\ g h_2 & 0 & 1 \end{pmatrix}}_{\mathbf{R}^{\frac{1}{2}}} \underbrace{\begin{pmatrix} z_3[k] \\ z_2[k] \\ z_2[k+1] \end{pmatrix}}_{\mathbf{z}}. \tag{3.132}$$

Again, the achievable rate follows by applying the results from [9], [10]:

$$R_2 = \frac{1}{2} \log \det \left(\mathbf{I}_2 + P\mathbf{R}^{-1} \mathcal{E} \left\{ \mathbf{z}\mathbf{z}^{\mathrm{H}} \right\} \mathbf{H}\mathbf{H}^{\mathrm{H}} \right). \tag{3.133}$$

$$= \frac{1}{2} \log \left(1 + \frac{P|h_0|^2}{\sigma_2^2} + \frac{P|h_1 g h_2|^2}{(|g h_2|^2 \sigma_3^2 + \sigma_2^2)} \right) \tag{3.134}$$

$$+ \frac{1}{2} \log \left(1 + \frac{P|h_0|^2}{P|h_2|^2 + \sigma_3^2} \right) \tag{3.135}$$

It can be seen directly from (3.127) and (3.135), that $R_2 > R_1$. However, since in protocol P2 the source transmits in every time slot with P it follows that the average energy consumption of protocol P2 over two time slots is twice the average energy consumption of protocol P1.

3.2.4 Numerical Examples

Full-duplex terminals. We again consider the network model depicted in Fig.3.8, but with Rayleigh fading coefficients $h_0 = c_0$, $h_1 = c_1/d^{\delta/2}$ and $h_2 = c_2/(1-d)^{\delta/2}$, where the $c_i \sim \mathcal{CN}(0, 1)$, $i = 0, 1, 2$ are independent of each other. We choose $P_1 = P_3 = 10$ and $\sigma_2^2 = \sigma_3^2 = 1$. We assume that in each symbol interval the channel gains $h_0.h_1, h_2$ change independently and that therefore each codeword captures enough channel fluctuations such that ergodic signaling is feasible [103]. Fig.3.11 plots the ergodic rates of the different relaying schemes. As main difference to the nonfading case the compress-and-forward scheme now surprisingly almost achieves the cut-set upper bound for $d = 1$ *and* $d = 0$, i.e., compress-and-forward is almost capacity achieving when the relay is near to the source or to the destination. The curve labeled with compress-and-forward II refers to a relay quantization rule without taking advantage of the available side information at the destination [84]. This approach is useful, when the relay does not know the bCSI of the source-destination channel, which is necessary to consider the side information for the quantization at the relay. Note however, that the compress-and-forward schemes need full CSI at all nodes.

Half-duplex terminals. We provide numerical examples for AWGN channels. For fading channels the same conclusions hold as for the full-duplex case, i.e., the optimum correlation coefficient is given by $\beta = 0$, due to lack of CSI at the transmitters. For all examples we again choose the network depicted in Fig.3.8.

In Fig.3.12 we compare the achievable rates of the half-duplex relaying schemes with respect to the cut-set upper bound and the direct communication between source and destination. Thereby we choose equal relay-receive and relay-transmit periods, i.e., $\alpha = 1/2$. The transmit power of each node is P, i.e., each node consumes the energy $P/2 \cdot T$ in one time slot. The pathloss coefficient is $\delta = 2$. We observe that as in the full-duplex case DF achieves the cut-set bound when the relay is near to the source and CF achieves the cut-set bound when the relay is near to the destination. Again the CF strategy is always better than direct communication whereas the DF rate drops below the rate of the direct communication when the source-relay SNR is below the source-destination SNR. We also observe the the AF protocol P1 is worse than the AF protocol P2, which corresponds to the result obtained in [104]. In contrast to the full-duplex case the amplify-and-forward protocols perform always worse than direct communication. This is because the pathloss coefficient $\delta = 2$ is too small for making AF more favorable than direct communication. Figures 3.13 and 3.14 show the performance of both AF protocols for pathloss coefficients $2, 4, 6$. Both protocols become better when the pathloss coefficient δ is increased.

In Fig. 3.15 we look at the achievable rates of the relaying schemes when the relay-receive and relay-transmit periods (adjusted by α) are optimized such that the rates are maximized. We also allow for power allocation: the source chooses power $P/(2 * \alpha)$ in the first time slot and $P/(2 * (1 - \alpha))$ in the second time slot. The relay transmits in the second time slot with power $P/(2 * (1 - \alpha))$. With that each node consume an energy of $P/2 \cdot T$ in one time slot (as in Fig.3.12. We observe that the cut-set bound and the DF rate are substantially increased when the relay is near to the source, because the relay is allowed to transmit with a very high power (since $1 - \alpha \to 0$ for $d \to 0$ boosting the multiple-access capacity (which is usually the bottleneck when the source-relay channel is strong). In Fig.3.16 we show the optimum relay-receive periods α for the cut-set bound, DF and CF rate. We observe that when the relay is near to the source, the cut-set bound and DF strategy are optimized when the relay gets almost the whole time slot for reception and only a very short time for transmission. The CF gets more time for reception when it moves towards the destination and less time when the relay moves towards the source.

In Fig.3.17 we look at the rates when the relay-receive period is optimized but no power allocation is allowed, i.e., each node transmits with power P irrespective of α. Surprisingly, the CF rate is not able to achieve the cut-set bound.

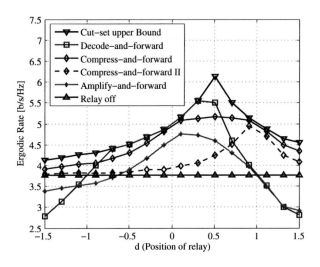

Fig. 3.11: Ergodic rates for Rayleigh fading. Compress-and-forward II relates to quantization without side information [84]

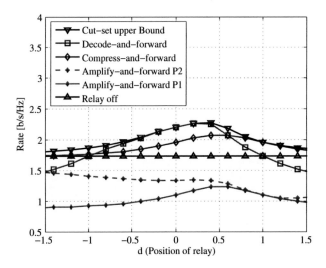

Fig. 3.12: Achievable rates for half-duplex AWGN relay channel with equal relay-receive and relay-transmit periods ($\alpha = \frac{1}{2}$)

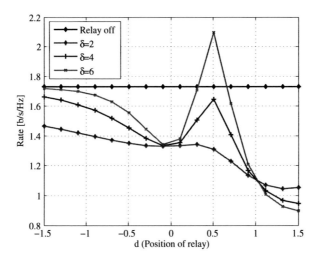

Fig. 3.13: Achievable rates of AF Protocol P2 for half-duplex AWGN relay channel with equal relay-receive and relay-transmit periods ($\alpha = \frac{1}{2}$) and different path losses

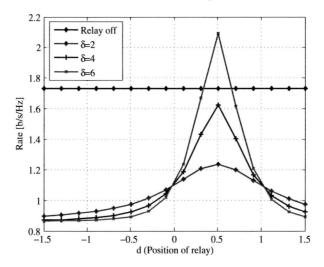

Fig. 3.14: Achievable rates of AF Protocol P1 for half-duplex AWGN relay channel with equal relay-receive and relay-transmit periods ($\alpha = \frac{1}{2}$) and different path losses

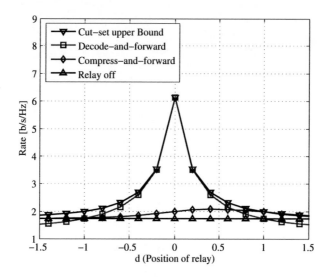

Fig. 3.15: Achievable rates for half-duplex AWGN relay channel with optimized relay-receive and relay-transmit periods and power allocation

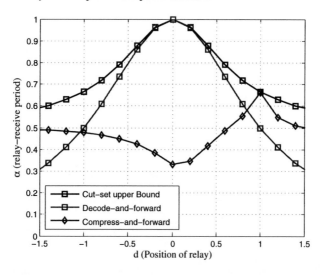

Fig. 3.16: Optimum relay-receive periods α for AWGN half-duplex relay channel

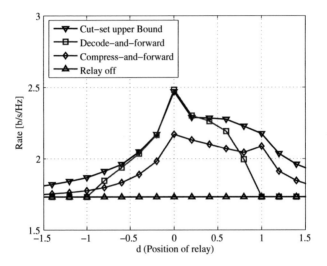

Fig. 3.17: Achievable rates for half-duplex AWGN relay channel with optimized relay-receive and relay-transmit periods and *no* power allocation

4 Achievable Rate Regions for the Two-way Relay Channel

In a two-way communication channel two terminals simultaneously transmit their messages to each other and the messages interfere with each other. This channel was first studied by Shannon [105] where he found an inner and an outer bound on the capacity region. It was shown that the inner bound coincides with the capacity region of the *restricted* two-way channel (TWC). In a restricted TWC the encoders of both terminals do not cooperate and the transmitted symbols at one terminal only depend on the message to be transmitted at that terminal and not on the past received symbols. In [106] it was further shown, that the capacity region of the Gaussian TWC coincides with the inner bound, i.e., the encoders do not need to cooperate in order to achieve the capacity region. The capacity region of the general TWC, i.e, when both encoders are allowed to cooperate via their received symbols, is still unknown.

In this chapter we consider the restricted two-way communication problem for the relay channel (TWRC), i.e., two terminals transmit their messages to each other whereas a third terminal helps both terminals in the communication process, see Fig.4.1. A motivation for considering the TWRC problem is found in wireless relaying where terminals are likely to operate in a half-duplex mode, i.e., terminals cannot transmit and receive at the same time in the same frequency band. Often the relaying strategies employed to such networks suffer from a pre-log factor $\frac{1}{2}$ in the corresponding expression for the capacity or achievable rate [27]. In the next chapter we will show that the half-duplex loss in spectral efficiency of relay networks can be recovered by two-way relaying. Further details are given in the next chapter.

In this chapter we focus on full-duplex two-way relay channels. As discussed in the previous chapter, there are several strategies known for one-way relay channels: decode-and-forward [14], [15], compress-and-forward [14], [15] or amplify-and-forward [27]. We are interested in how these strategies can be extended to the two-way relay channel with full-duplex terminals and which rate regions are achievable.

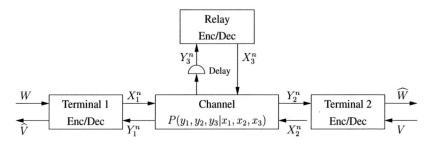

Fig. 4.1: Two-way Relay Channel

4.1 System Model

We consider a two-way relay channel (TWRC) with three full-duplex terminals. Terminals T_1 and T_2 want to exchange messages (two-way communication) with the help of terminal T_3 which acts as relay and has no own messages to transmit. The random variables of the channel are:

- the forward message $W \in \mathcal{W}$ of terminal T_1,

- the backward message $V \in \mathcal{V}$ of terminal T_2,

- the channel inputs (transmitted symbols) $X_{ti} \in \mathcal{X}$, $t = 1, 2, 3$, $i = 1, 2, \ldots, n$,

- the channel outputs (received symbols) $Y_{ti} \in \mathcal{Y}$, $t = 1, 2, 3$, $i = 1, 2, \ldots, n$,

- the message estimate $\widehat{W} \in \mathcal{W}$ at terminal T_2,

- the message estimate $\widehat{V} \in \mathcal{V}$ at terminal T_1,

where $\mathcal{W} = \{1, 2, \ldots, 2^{B_W}\}$ and $\mathcal{V} = \{1, 2, \ldots, 2^{B_V}\}$, i.e., message W carries B_W information bits and message V carries B_V information bits. The input alphabet is given by $\mathcal{X}$ and the output alphabet by $\mathcal{Y}$[1]. For the *general* TWRC the transmit symbol of terminal T_1 at the discrete time k is a function of the message w to be transmitted and its past channel outputs $y_1^{k-1} = y_{11}, y_{12}, \ldots, y_{1(k-1)}$. Encoding at terminal T_1 is therefore defined by the sequence of encoding functions

$$f_{11}(w), f_{12}(w, y_1^1), f_{13}(w, y_1^2), \cdots, f_{1n}(w, y_1^{n-1}). \tag{4.1}$$

---

[1]We assume the same alphabet for all channel inputs and outputs.

Similarly the transmit symbol of terminal T_2 at the discrete time k is a function of the message v to be transmitted and its past channel outputs $y_2^{n-1} = y_{21}, y_{22}, \ldots, y_{2(k-1)}$, i.e.,

$$f_{21}(v), f_{22}(v, y_2^1), f_{23}(v, y_2^2), \cdots, f_{2n}(v, y_2^{n-1}). \tag{4.2}$$

In this work we consider the *restricted* TWRC [105] where the channel inputs at terminals T_1 and T_2 are functions of the messages to be transmitted only, i.e.,

$$x_1^n = f_1(w) \tag{4.3}$$
$$x_2^n = f_2(v). \tag{4.4}$$

for all $i = 1, 2, \ldots, n$. The rate regions obtained for the restricted TWRC serve as inner bounds for the capacity region of the general TWRC. The transmit symbols of relay terminal T_3 are determined by the relay's past channel outputs, i.e.,

$$x_{3i} = f_{3i}(y_3^{i-1}) \tag{4.5}$$

for $i = 1, 2, \ldots, n$. We assume a time invariant and memoryless TWRC which is defined by the conditional channel distribution

$$p_{Y_1 Y_2 Y_3 | X_1 X_2 X_3}(y_1, y_2, y_3 | x_1, x_2, x_3) \tag{4.6}$$

where X_t and Y_t, $t = 1, 2, 3$ are random variables representing the respective channel inputs and outputs[2]. Terminal T_1 computes its message estimate $\widehat{V}$ as a function of its past channel outputs Y_1^n and its transmitted message W

$$\widehat{v} = g_1(w, y_1^n) \tag{4.7}$$

whereas terminal T_2 computes its message estimate $\widehat{W}$ as a function of its past channel outputs Y_2^n and its transmitted message V

$$\widehat{w} = g_2(v, y_2^n). \tag{4.8}$$

---

[2]Again: subscripts in the probability distributions will be dropped when they are obvious by inspection of the arguments.

The pair (R_1, R_2) of nonnegative numbers is called an achievable rate pair if for any $\delta > 0, \epsilon > 0$ and sufficiently large n there exist encoding and decoding functions such that

$$\frac{B_W}{n} \geq R_1 - \delta \tag{4.9}$$

$$\frac{B_V}{n} \geq R_2 - \delta \tag{4.10}$$

with

$$\Pr(\widehat{W} \neq W) \leq \epsilon \tag{4.11}$$

$$\Pr(\widehat{V} \neq V) \leq \epsilon. \tag{4.12}$$

The set of all rate pairs (R_1, R_2) is called the achievable rate region of the TWRC.

4.2 Two-way Decode-and-forward

At terminal T_1 the message w is divided into K blocks $w_1, w_2, \ldots, w_K$ of nR_1 bits each. The transmission is performed in $K + 1$ blocks, i.e., $B_W = KnR_1$ message bits are transmitted in $N = (K + 1)n$ channel uses with an overall rate $R_1 K/(K + 1)$. The codeword of length n at terminal T_1 consists of a superposition of two codewords:

$$x_1^n(i, j) = \sqrt{\alpha_1} x_{3f}^n(i) + \sqrt{\beta_1} \widetilde{x}_1^n(j) \tag{4.13}$$

where i and j range from 1 to 2^{nR_1} and α_1, β_1 are scaling coefficients. The $n \cdot 2^{nR_1}$ symbols $\widetilde{x}_{1m}(j)$ are chosen independently according to $P_{\widetilde{X}_1}(\cdot)$ and the $n \cdot 2^{nR_1}$ symbols $x_{3fm}(i)$ are chosen independently according to $P_{X_{3f}}(\cdot)$. The same procedure is performed at terminal T_2: the message v is divided into K blocks $v_1, v_2, \ldots, v_K$ of nR_2 bits each and transmitted in $K + 1$ blocks. The codeword at terminal T_2 consists of two codewords:

$$x_2^n(p, q) = \sqrt{\alpha_2} x_{3b}^n(p) + \sqrt{\beta_2} \widetilde{x}_2^n(q) \tag{4.14}$$

where p and q range from 1 to 2^{nR_2} and α_2, β_2 are scaling coefficients. The $n \cdot 2^{nR_2}$ symbols $\widetilde{x}_{2m}(q)$ are chosen independently according to $P_{\widetilde{X}_2}(\cdot)$ and the $n \cdot 2^{nR_2}$ symbols $x_{3bm}(p)$ are chosen independently according to $P_{X_{3b}}(\cdot)$. Note that for $K \to \infty$ and for fixed n the rates $R_1 K/(K + 1)$ and $R_2 K/(K + 1)$ are arbitrary close to R_1 and R_2, respectively.

Block Markov Superposition Coding. In the first block, $k = 1$, terminals T_1 and T_2 transmit

$x_1(1, w_1)$	$x_1(w_1, w_2)$	$x_1(w_2, w_3)$	$x_1(w_3, 1)$
$x_2(1, v_1)$	$x_2(v_1, v_2)$	$x_2(v_2, v_3)$	$x_2(v_3, 1)$
$x_3(1, 1)$	$x_3(w_1, v_1)$	$x_3(w_2, v_2)$	$x_3(w_3, v_3)$

Fig. 4.2: Two-way Block Markov encoding for $K = 4$

$x_1^n(1, w_1)$ and $x_2^n(1, v_1)$, respectively. The relay terminal T_3 transmits $x_3^n(1, 1) = x_{3f}^n(1) + x_{3b}^n(1)$, where $x_{3f}^n(1)$ denotes the codeword associated to the *forward* direction $T_1 \rightarrow T_2$ and $x_{3b}^n(1)$ the codeword associated to the *backward* direction $T_1 \leftarrow T_2$. Terminal T_3 is able to decode w_1 and v_1 as long as

$$R_1 \leq I(X_1; Y_3 | X_2 X_3) \tag{4.15}$$

$$R_2 \leq I(X_2; Y_3 | X_1 X_3) \tag{4.16}$$

$$R_1 + R_2 \leq I(X_1 X_2; Y_3 | X_3) \tag{4.17}$$

and the block size n is large. (4.15)–(4.17) corresponds to the rate region of a *multiple access channel*. The additional conditioning on X_3 is because the relay transmits at the same time a codeword x_3^n, but the relay is able to cancel this influence perfectly when decoding x_1^n and x_2^{n3}. In the second block, $k = 2$, terminals T_1 and T_2 transmit $x_1^n(w_1, w_2)$ and $x_2^n(v_1, v_2)$, respectively, and in the same time relay terminal T_3 transmits

$$x_3^n(w_1, v_1) = x_{3f}^n(w_1) + x_{3b}^n(v_1). \tag{4.18}$$

Terminal T_3 is able to decode w_2 and v_2 as long as n is large and (4.15)–(4.17) are true. One continues in this way until block $K + 1$, where terminals T_1 and T_2 transmit $x_1^n(w_K, 1)$ and $x_2^n(v_K, 1)$, respectively, and where terminal T_3 transmits

$$x_3^n(w_K, v_K) = x_{3f}^n(w_K) + x_{3b}^n(v_K). \tag{4.19}$$

The transmission scheme is depicted in Fig.4.2.

Sliding Window Decoding. Let y_{2k}^n and $y_{2(k-1)}^n$ be the observed symbols at terminal T_1 and T_2 in block k and $k - 1$, respectively. Starting right after block k, terminal T_2 decodes w_k from

---

[3]If the conditioning on X_3 would not appear in (4.15)–(4.17), this would model a situation where x_3^n interferes with the reception of x_1^n and x_2^n.

y_{2k}^n and $y_{2(k-1)}^n$ as long as

$$R \leq I(X_3; Y_2|X_2) + I(X_1; Y_2|X_2X_3) \tag{4.20}$$

$$= I(X_1X_3; Y_2|X_2) \tag{4.21}$$

and the block size n is large. This is analogues to the one-way decode-and-forward scheme. The reason is, that terminal T_2 knows v_k (back-propagating self-interference) and can therefore subtract this influence when decoding the message from terminal T_1. After decoding w_k the decoder shifts the window by one block and decodes w_{k+1} from $y_{2(k+1)}^n$ and y_{2k}^n. Since w_k is already decoded, it's influence on y_{2k}^n can also be subtracted. The message w_{k+1} can be decoded again as long as (4.21) is fulfilled and n is large. One continues this way until block $k = K$ where finally all messages $w_1, \ldots, w_K$ are decoded.

Similarly, terminal T_1 knows w_k (back-propagating self-interference at terminal T_1) and decodes v_k from y_{1k}^n and $y_{1(k-1)}^n$ as long as

$$R_2 \leq I(X_2X_3; Y_1|X_1). \tag{4.22}$$

After decoding v_k the decoder shifts the window by one block and decodes v_{k+1} from $y_{1(k+1)}^n$ and y_{1k}^n. One continues this way until block $k = K$ where finally all messages $v_1, \ldots, v_K$ are decoded. Combining (4.15)–(4.17) with (4.21) and (4.22) we obtain the achievable rate region for this scheme in the following proposition.

Proposition 4.2.1 (Two-way decode-and-forward). *An achievable rate region of the TWRC is given by* $\bigcup \{R_1, R_2\}$ *with*

$$R_1 \leq \min\left(I(X_1; Y_3|X_2X_3), I(X_1X_3; Y_2|X_2)\right) \tag{4.23}$$

$$R_2 \leq \min\left(I(X_2; Y_3|X_1X_3), I(X_2X_3; Y_1|X_1)\right) \tag{4.24}$$

$$R_1 + R_2 \leq I(X_1X_2; Y_3|X_3) \tag{4.25}$$

where the union is over all product distributions $p(x_1|x_{3f})p(x_2|x_{3b})p(x_{3f})p(x_{3b})$.

Next, we establish the rate region of the memoryless additive white Gaussian noise TWRC. The received symbols at terminals T_1, T_2 and T_3 are defined by

$$Y_1 = h_0X_2 + h_1X_3 + Z_1 \tag{4.26}$$

$$Y_2 = h_0X_1 + h_2X_3 + Z_2 \tag{4.27}$$

$$Y_3 = h_1X_1 + h_2X_2 + Z_3 \tag{4.28}$$

where $h_0, h_1, h_2 \in \mathbb{C}$ denote deterministic channel gains[4] between terminals T_1 and T_2, terminals T_1 and T_3, and terminals T_2 and T_3, respectively. The $Z_i \in \mathbb{C}$ denote independent, complex, proper, Gaussian random variables with zero mean and unit variances. All input and output symbols are taken from the complex alphabet $\mathbb{C}$. We impose the per-symbol power constraints

$$\frac{1}{n} \sum_{m=1}^{n} |x_{3fm}(i)|^2 \leq \gamma P_3 \tag{4.29}$$

$$\frac{1}{n} \sum_{m=1}^{n} |x_{3bm}(q)|^2 \leq \overline{\gamma} P_3 \tag{4.30}$$

$$\frac{1}{n} \sum_{m=1}^{n} |\widetilde{x}_{1m}(j)|^2 \leq P_1 \tag{4.31}$$

$$\frac{1}{n} \sum_{m=1}^{n} |\widetilde{x}_{2m}(q)|^2 \leq P_2 \tag{4.32}$$

for all $i, j \in [1, 2^{nR_1}]$, $p, q \in [1, 2^{nR_2}]$ and $\overline{\gamma} = (1 - \gamma)$. Let ρ_1 be the complex correlation coefficient between codeword x_1^n and x_3^n and ρ_2 the complex correlation coefficient between codeword x_2^n and x_3^n, defined as

$$\rho_1 = \frac{\mathcal{E}\{X_1 X_3^*\}}{\sqrt{P_1 P_3}} \tag{4.33}$$

$$\rho_2 = \frac{\mathcal{E}\{X_2 X_3^*\}}{\sqrt{P_2 P_3}}. \tag{4.34}$$

The scaling coefficients α_1 and β_1 for the codeword of terminal T_1, see (4.13), are chosen as

$$\alpha_1 = \frac{|\rho_1|^2 P_1}{\gamma P_3} \tag{4.35}$$

$$\beta_1 = 1 - |\rho_1|^2. \tag{4.36}$$

The scaling coefficients α_2 and β_2 for the codeword of terminal T_2, see (4.14), are correspondingly chosen as

$$\alpha_2 = \frac{|\rho_2|^2 P_2}{\overline{\gamma} P_3} \tag{4.37}$$

$$\beta_2 = 1 - |\rho_2|^2. \tag{4.38}$$

These scaling coefficients ensure that the average power of codeword x_1^n is P_1 and of x_2^n is

---

[4]We assume reciprocity for all channel gains.

P_2. The power of x_3^n is P_3, whereas γP_3 is used for the *forward* codeword x_{3f}^n and $\overline{\gamma} P_3$ for the *backward* codeword x_{3b}^n. It remains now to evaluate (4.23)–(4.25). The mutual information between source and relay follows as

$$I(X_1; Y_3|X_2X_3) = h(Y_3|X_2X_3) - h(Y_3|X_1X_2X_3) \tag{4.39}$$

$$= h(Y_3|X_2X_3) - h(Z_3) \tag{4.40}$$

$$= h(Y_3, X_3|X_2) - h(X_3|X_2) - h(Z_3) \tag{4.41}$$

$$= \log\left((\pi e)^2 \left|R_{Y_3X_3|X_2}\right|^2\right) - \log\left(\pi e \gamma P_3\right) - \log\left(\pi e \sigma_3^3\right). \tag{4.42}$$

The covariance matrix $R_{Y_3X_3|X_2}$ follows as

$$R_{Y_3X_3|X_2} = \begin{pmatrix} P_1|h_1|^2 + \sigma_3^2 & h_1\mathcal{E}\left\{X_1X_3^*|X_2\right\} \\ h_1^*\mathcal{E}\left\{X_1^*X_3|X_2\right\} & \gamma P_3 \end{pmatrix} \tag{4.43}$$

Inserting the determinant of (4.43) into (4.39) leads to

$$I(X_1; Y_3|X_2X_3) = \log\left(1 + \frac{P_1|h_1|^2\left(1 - |\rho_1|^2\right)}{\sigma_3^3}\right) \tag{4.44}$$

The same considerations lead to the corresponding mutual information in the backward direction:

$$I(X_2; Y_3|X_1X_3) = \log\left(1 + \frac{P_2|h_2|^2\left(1 - |\rho_2|^2\right)}{\sigma_3^3}\right) \tag{4.45}$$

We now evaluate the mutual information of the multiple access cut in the forward direction:

$$I(X_1X_3; Y_2|X_2) = h(Y_2|X_2) - h(Y_2|X_1X_2X_3) \tag{4.46}$$

$$= h(X_1h_0 + X_3h_2 + Z_2|X_2) - h(Z_2) \tag{4.47}$$

$$= \log\left(\pi e \cdot \mathcal{V}\left\{X_1h_0 + X_3h_2 + Z_2|X_2\right\}\right) - \log\left(\pi e \sigma_2^2\right) \tag{4.48}$$

$$= \log\left(1 + \frac{P_1|h_0|^2 + \gamma P_3|h_2|^2 + 2\mathfrak{Re}\left\{h_0h_2^*\rho_1\right\}\sqrt{\gamma P_1 P_3}}{\sigma_2^2}\right) \tag{4.49}$$

where $\mathcal{V}\left\{x\right\}$ denotes the variance of x. The same considerations lead to the corresponding multiple access cut in the backward direction:

$$I(X_2X_3; Y_1|X_1) = \log\left(1 + \frac{P_2|h_0|^2 + P_3|h_1|^2 + 2\mathfrak{Re}\left\{h_0h_1^*\rho_2\right\}\sqrt{P_2 P_3}}{\sigma_1^2}\right). \tag{4.50}$$

By summarizing the results and setting $\sigma_1^2 = \sigma_2^2 = \sigma_3^3 = 1$ we obtain the rate region for the Gaussian TWRC in the following Proposition.

Proposition 4.2.2 (Two-way decode-and-forward for AWGN Channels).

$$\bigcup_{0 \leq \rho_1, \rho_2, \gamma \leq 1} \left\{ R_1, R_2 : \right.$$

$$R_1 \leq \min \left(C \left(P_1 |h_1|^2 \left(1 - |\rho_1|^2 \right) \right), C \left(P_1 |h_0|^2 + \gamma P_3 |h_2|^2 + 2 \Re \left\{ h_0 h_2^* \rho_1 \right\} \sqrt{\gamma P_1 P_3} \right) \right)$$

$$R_2 \leq \min \left(C \left(P_2 |h_2|^2 \left(1 - |\rho_2|^2 \right) \right), C \left(P_2 |h_0|^2 + \bar{\gamma} P_3 |h_1|^2 + 2 \Re \left\{ h_0 h_1^* \rho_2 \right\} \sqrt{\bar{\gamma} P_2 P_3} \right) \right)$$

$$\left. R_1 + R_2 \leq C \left(P_1 |h_1|^2 \left(1 - |\rho_1|^2 \right) + P_2 |h_2|^2 \left(1 - |\rho_2|^2 \right) \right) \right\}$$

with $C(x) = \log_2(1 + x)$. When γ is zero, the relay is not used for the forward direction, but terminal T_1 can still communicate with terminal T_2 over the direct link with the rate being $C(P_1 |h_0|^2)$. Similarly for $\gamma = 1$, the relay is used only for the forward direction, but not for the backward direction. The rate for the backward direction is then given by $C(P_2 |h_0|^2)$.

4.3 Two-way Compress-and-forward

In the previous section the relay was enforced to decode the symbols from both terminals. Another scheme where the relay does not decode but sends estimates of its received symbols was developed in [14, Theorem 6] for the one-way relay channel and is nowadays known as *compress-and-forward*, see the previous chapter or [14], [15]. We extend this scheme to the TWRC to obtain the following rate region.

Proposition 4.3.1 (Two-way compress-and-forward). *An achievable rate region of the TWRC is given by* $\bigcup \{ R_1, R_2 \}$ *with*

$$R_1 \leq I(X_1; Y_2 \widehat{Y}_3 | X_2 X_3) \tag{4.51}$$

$$R_2 \leq I(X_2; Y_1 \widehat{Y}_3 | X_1 X_3) \tag{4.52}$$

subject to the constraint

$$\max \left(I(\widehat{Y}_3; Y_3 | X_1 X_3 Y_1), I(\widehat{Y}_3; Y_3 | X_2 X_3 Y_2) \right)$$
$$< \min \left(I(X_3; Y_1 | X_1), I(X_3; Y_2 | X_2) \right) \tag{4.53}$$

where the union is over all distributions

$$p(x_1)p(x_2)p(x_3)p(y_1, y_2, y_3|x_1, x_2, x_3)p(\widehat{y}_3|x_3, y_3). \tag{4.54}$$

The auxiliary random variable $\widehat{Y}_3$ represents a quantized and compressed version of Y_3 that is available at terminals T_1 and T_2 since the constraint in (4.53) ensures that the index of the quantized codeword $\widehat{Y}_3^n$ is transmitted reliably to *both* terminals T_1 and T_2.

Proof of Proposition 4.3.1. The proof is based on the use of *typical sequences* and is along the lines of the proof of the one-way strategy [14].

Construction of Code Books. Terminal T_1 chooses 2^{nR_1} i.i.d. x_1^n each with probability

$$P(x_1^n) = \prod_{i=1}^{n} P(x_{1i}) \tag{4.55}$$

and labels the codewords by $w \in [1, 2^{nR_1}]$. Terminal T_2 chooses 2^{nR_2} i.i.d. x_2^n each with probability

$$P(x_2^n) = \prod_{i=1}^{n} P(x_{2i}) \tag{4.56}$$

and labels the codewords by $v \in [1, 2^{nR_2}]$. The relay terminal T_3 chooses 2^{nR_3} i.i.d. x_3^n each with probability

$$P(x_3^n) = \prod_{i=1}^{n} P(x_{3i}) \tag{4.57}$$

and labels the codewords by $s \in [1, 2^{nR_3}]$. The relay chooses then, for each $x_3^n(s)$, $2^{n(R_3+R_3')}$ i.i.d. $\widehat{Y}_3^n$ each with probability

$$P(\widehat{y}_3^n|x_3^n(s)) = \prod_{i=1}^{n} P(\widehat{y}_{3i}|x_{3i}(s)) \tag{4.58}$$

with

$$P(\widehat{y}_3|x_3) = \sum_{x_1,x_2,y_1,y_2,y_3} P(x_1)P(x_2)P(y_1, y_2, y_3|x_1, x_2, x_3)P(\widehat{y}_3|y_3, x_3) \tag{4.59}$$

where $P(x_1)$ denotes the distribution of the symbols of terminal T_1 and the $P(x_2)$ the distribution of the symbols of terminal T_2. The conditional channel distribution, i.e., the distribution of the channel outputs given the channel inputs is given by $P(y_1, y_2, y_3|x_1, x_2, x_3)$ and $P(\widehat{y}_3|y_3, x_3)$ is

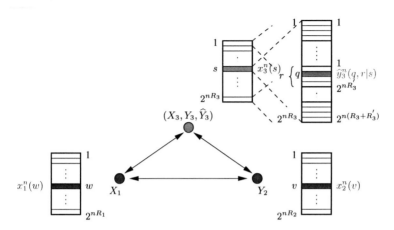

Fig. 4.3: Code book constructions for two-way compress-and-forward

the "quantization rule" the relay has to choose. Note that the channel inputs X_1 and X_2 are independent because we consider the restricted two-way relay channel. In general, these random variables may be dependent [105].

For a given s the relay labels then $\widehat{y}_3^n(q, r|s)$ by

- $q \in [1, 2^{nR_3'}]$ (index within bin)

- $r \in [1, 2^{nR_3}]$ (bin index)

see Fig.4.3. The auxiliary random variable $\widehat{Y}_3$ represents a quantized version of Y_3 and the joint distribution of the random variables $X_1, X_2, X_3, Y_1, Y_2, Y_3, \widehat{Y}_3$ factors as

$$P(x_1)P(x_2)P(x_3)P(y_1, y_2, y_3|x_1, x_2, x_3)P(\widehat{y}_3|x_3, y_3). \qquad (4.60)$$

Encoding at Terminal T_1. At terminal T_1 the message w of nR_1K bits is divided into K equally-sized blocks $w_1, w_2, \ldots, w_K$ of nR_1 bits each. In block k, $k = 1, 2, \ldots, K + 1$ terminal T_1 transmits $x_1^n(w_k)$ where $w_{K+1} = 1$.

Encoding at Terminal T_2. At terminal T_2 the message v of nR_2K bits is divided into K equally-sized blocks $v_1, v_2, \ldots, v_K$ of nR_2 bits each. In block k, $k = 1, 2, \ldots, K + 1$ terminal T_2 transmits $x_2^n(v_k)$ where $v_{K+1} = 1$.

Encoding at the Relay. After block k the relay terminal T_3 observes y_{3k}^n and tries to find a

(q_k, r_k) from the quantization code book that is indexed by $s_k = r_{k-1}$ such that

$$\left(\widehat{y}_3^n(q_k, r_k | s_k = r_{k-1}), y_{3k}^n, x_3^n(s_k = r_{k-1})\right) \in T_\epsilon^n \left(P_{\widehat{Y}_3 Y_3 X_3}\right) \tag{4.61}$$

where $T_\epsilon^n \left(P_{\widehat{Y}_3 Y_3 X_3}\right)$ denotes the jointly typical set with respect to $P_{\widehat{Y}_3 Y_3 X_3}$. If one or more such (q_k, r_k) are found the relay chooses one of them, sets $s_{k+1} = r_k$ and transmits $x_3(s_{k+1} = r_k)$ in block $k + 1$, i.e., the relay transmits a codeword in block $k + 1$ that represents the bin index of the quantized relay observation in block k. If no such pair is found the relay sets $s_{k+1} = 1$ and transmits $x_3(1)$. Notice that in block $k = 1$ the relay sets $s_1 = 1$. We know from rate distortion theory that the relay is able to find a (q_k, r_k) with high probability when

$$R_3 + R_3' > I(\widehat{Y}_3; Y_3 | X_3) \tag{4.62}$$

and the block size n is large. Relation (4.62) follows from the conditional joint typicality bound given in Theorem 2.1.4 by setting $u^n = x_3^n(s_k)$, $X^n = \widehat{Y}_3^n(q_k, r_k | s_k)$ and $y^n = y_{3k}^n$:

$$\Pr\left[\left(\widehat{Y}_3^n(q_k, r_k | s_k), y_{3k}^n, x_3^n(s_k)\right) \in T_\epsilon^n \left(P_{\widehat{Y}_3 Y_3 X_3}\right)\right] \geq (1 - \epsilon_2'(n)) 2^{-n(I(\widehat{Y}_3; Y_3 | X_3) + \delta)} \tag{4.63}$$

where $\delta \to 0$ when $\epsilon \to 0$. Suppose now that $(y_{3k}^n, x_3^n) \in T_\epsilon^n (P_{Y_3 X_3})^5$ but none of the $\widehat{y}_3^n$ satisfies $(y_{3k}^n, x_3^n, \widehat{y}_3^n) \in T_\epsilon^n (P_{Y_3 X_3 \widehat{Y}_3})$. The probability of this event is independent for each (q_k, r_k) due to the code book construction (4.58). The probability $P_e^n(\widehat{y}_3^n)$ that none of the codewords $\widehat{y}_3^n$ are satisfactory is thus

$$P_e^n(\widehat{y}_3^n) = \left(1 - \Pr\left[\left(\widehat{Y}_3^n(q_k, r_k | s_k), y_{3k}^n, x_3^n(s_k)\right) \in T_\epsilon^n \left(P_{\widehat{Y}_3 Y_3 X_3}\right)\right]\right)^{2^{n(R_3 + R_3')}} \tag{4.64}$$

$$\leq \exp\left(-2^{n(R_3 + R_3' - I(\widehat{Y}_3; Y_3 | X_3) + \delta)}\right) \tag{4.65}$$

where we used the inequality $(1 - x)^n \leq \exp(-nx)$. It follows that we must choose $R_3 + R_3' > I(\widehat{Y}_3; Y_3 | X_3)$ for the error probability $P_e^n(\widehat{y}_3^n)$ to tend to zero as $n \to \infty$. If two or more codewords $\widehat{y}_3^n$ satisfy the typicality check (4.61), the relay can choose one of them because every quantization $\widehat{y}_3^n$ that satisfies the typicality check is a good representation of the relay observation y_3^n and the ambiguity does not lead to errors. Roughly spoken, the relation (4.62) says that the relay should not compress y_3^n too much, otherwise the relay may not be able to find an appropriate quantization codeword that satisfies (4.61).

---

[5]The probability of this event approaches one for $n \to \infty$ as can be seen from the extension of (2.4) to joint typicality [74].

Decoding at Terminals T_1 **and** T_2. After block k terminal T_2 tries to estimate s_k by looking for a $\tilde{s}_k$ such that

$$(x_3^n(\tilde{s}_k), y_{2k}^n) \in T_\epsilon^n \left(P_{X_3 Y_2 | X_2} \right) \tag{4.66}$$

and if successful puts out $r_{k-1} = \tilde{s}_k$. Terminal T_1 does the same, i.e., looks for $\tilde{s}_k$ such that

$$(x_3^n(\tilde{s}_k), y_{1k}^n) \in T_\epsilon^n \left(P_{X_3 Y_1 | X_1} \right) \tag{4.67}$$

and if successful puts out $r_{k-1} = \tilde{s}_k$. Conditioning on X_i, $i = 1, 2$ is because both terminals simultaneously transmit during they receive (full-duplex) and the self-interference of for example x_2^n on y_2^n is considered to be canceled perfectly. Both terminals succeed with high probability if

$$R_3 < \min(I(X_3; Y_1 | X_1), I(X_3; Y_2 | X_2)) \tag{4.68}$$

and the block size n is large. Relation (4.68) follows by the application of Theorems 2.1.2 and 2.1.3. For the decoder at terminal T_2:

$$\Pr\left[(X_3^n, y_2^n) \in T_\epsilon^n \left(P_{X_3 Y_2 | X_2} \right) \right] \leq 2^{-n(I(X_3; Y_2 | X_2) - \delta)}. \tag{4.69}$$

Now suppose that $(x_3^n(s_k), y_{2k}^n) \in T_\epsilon^n \left(P_{X_3 Y_2 | X_2} \right)^6$ but that the decoder finds more than one x_3^n that is jointly typical with the received sequence. The probability for this event is

$$P_e^n(x_3^n) = \Pr\left[\bigcup_{\tilde{s} \neq s} (X_3^n(\tilde{s}), y_2^n) \in T_\epsilon^n \left(P_{X_3 Y_2 | X_2} \right) \right] \tag{4.70}$$

$$\leq \sum_{\tilde{s} \neq s} 2^{-n(I(X_3; Y_2 | X_2) - \delta)} \tag{4.71}$$

$$\leq 2^{nR_3} 2^{-n(I(X_3; Y_2 | X_2) - \delta)} \tag{4.72}$$

$$= 2^{n(R_3 - I(X_3; Y_2 | X_2) + \delta)} \tag{4.73}$$

where the first inequality follows by the union bound for probabilities. The same conclusions follow for the decoder at terminal T_1:

$$\Pr\left[(X_3^n, y_1^n) \in T_\epsilon^n \left(P_{X_3 Y_1 | X_1} \right) \right] \leq 2^{-n(I(X_3; Y_1 | X_1) - \delta)}. \tag{4.74}$$

Again, for $(x_3^n(s_k), y_{1k}^n) \in T_\epsilon^n \left(P_{X_3 Y_1 | X_1} \right)$ it is possible that the decoder finds more than one x_3^n

---

[6]Again, by joint typicality the probability of this event approaches one for $n \to \infty$.

that is jointly typical with the received sequence. The probability for this event is

$$P_e^n(x_3^n) \leq 2^{n(R_3 - I(X_3; Y_1 | X_1) + \delta)}. \tag{4.75}$$

In order to drive the probability to zero that one of the terminals does not succeed in decoding the bin index one should choose a large n and the rate R_3 as in (4.68). Relation (4.68) says, that the bin index $s_k = r_{k-1}$ (*coarse* information about the quantized relay observation) has to be communicated reliably to *both* terminals T_1 and T_2.

Afterwards, terminal T_2 uses $y_{2(k-1)}^n$ to find a $\widetilde{q}_{k-1}$ (*fine* information about the quantized relay observation) such that

$$(\widehat{y}_3^n(\widetilde{q}_{k-1}, r_{k-1} | s_{k-1}), y_{2(k-1)}^n, x_3^n(s_{k-1})) \in T_\epsilon^n \left(P_{\widehat{Y}_3 Y_2 X_3 | X_2} \right) \tag{4.76}$$

and if successful puts out $q_{k-1} = \widetilde{q}_{k-1}$. Terminal T_2 succeeds with high probability when

$$R_3' < I(\widehat{Y}_3; Y_2 | X_2 X_3) \tag{4.77}$$

and the block size n is large. The Markov Lemma 2.1.5 ensures, that the probability of (4.76) goes to one as (4.77) is fulfilled and n grows large. The Markov Lemma says that if $\widehat{Y}_3$ and Y_2 are jointly typical and $\widehat{Y}_3$ and X_3 are jointly typical so are Y_2 and X_3. Another source of error is when the decoder finds several $\widetilde{\widetilde{q}}_{k-1}$ that satisfy the typicality check.

$$(\widehat{y}_3^n(\widetilde{\widetilde{q}}_{k-1}, r_{k-1} | s_{k-1}), y_{2(k-1)}^n, x_3^n(s_{k-1})) \in T_\epsilon^n \left(P_{\widehat{Y}_3 Y_2 X_3 | X_2} \right). \tag{4.78}$$

The probability for this event is upper bounded by

$$P_e^n(\widehat{y}_3^n) = \Pr\left[\bigcup_{\widetilde{\widetilde{q}}_{k-1} \neq \widetilde{q}_{k-1}} \left(\widehat{Y}_3^n(\widetilde{\widetilde{q}}_{k-1}, r_{k-1} | s_{k-1}), y_{2(k-1)}^n, x_3^n(s_{k-1}) \right) \right.$$

$$\left. \in T_\epsilon^n \left(P_{\widehat{Y}_3 Y_2 X_3 | X_2} \right) \right] \tag{4.79}$$

$$\leq \sum_{\widetilde{\widetilde{q}}_{k-1} \neq \widetilde{q}_{k-1}} 2^{-n(I(\widehat{Y}_3; Y_2 | X_2 X_3) - \delta)} \tag{4.80}$$

$$< 2^{nR_3'} 2^{-n(I(\widehat{Y}_3; Y_2 | X_2 X_3) - \delta)} \tag{4.81}$$

$$= 2^{n(R_3' - I(\widehat{Y}_3; Y_2 | X_2 X_3) - \delta)} \tag{4.82}$$

and from that we get (4.77). This means that there shouldn't be too much codewords in one bin, otherwise the decoder at terminal T_2 is not able to resolve the uncertainty (to decode the *fine* information r) with help of the side information Y_2. The same conclusions hold for terminal T_1, one has simply to replace $y_{2(k-1)}^n$ by $y_{1(k-1)}^n$. Therefore, both terminals succeed with high probability when

$$R_3' < \min(I(\widehat{Y}_3; Y_1 | X_1 X_3), I(\widehat{Y}_3; Y_2 | X_2 X_3)) \tag{4.83}$$

and the block size n is large. Since the quantization has to work for both terminals, the number $2^{nR_3'}$ of codewords per bin is chosen such, that the weaker terminal may decode the quantized codeword $\widehat{Y}_3^n$.

Finally, terminal T_2 uses both $y_{2(k-1)}^n$ and $\widehat{y}_3^n(q_{k-1}, r_{k-1} | s_{k-1})$ to find an index $\widetilde{w}_{k-1}$ such that

$$(x_1^n(\widehat{w}_{k-1}), \widehat{y}_3(q_{k-1}, r_{k-1} | s_{k-1}), y_{2(k-1)}^n, x_3^n(s_{k-1}) | x_2^n(v_{k-2})) \in T_\epsilon^n\left(P_{X_1 \widehat{Y}_3 Y_2 X_3 | X_2}\right) \tag{4.84}$$

and if successful puts out $w_{k-1} = \widehat{w}_{k-2}$ as result. Terminal T_2 succeeds with high probability when

$$R_1 < I(X_1; Y_2 \widehat{Y}_3 | X_2 X_3) \tag{4.85}$$

and the block size n is large which follows by the same argumentation as for (4.68) and (4.83). Similar, terminal T_1 succeeds with high probability in decoding v_{k-1} when

$$R_2 < I(X_2; Y_1 \widehat{Y}_3 | X_1 X_3). \tag{4.86}$$

By choosing

$$R_3' = \min\left(I(\widehat{Y}_3; Y_1 | X_1 X_3), I(\widehat{Y}_3; Y_2 | X_2 X_3)\right) - \epsilon \tag{4.87}$$

we get

$$R_3 > I(\widehat{Y}_3; Y_3 | X_3) - \min\left(I(\widehat{Y}_3; Y_1 | X_1 X_3), I(\widehat{Y}_3; Y_2 | X_2 X_3)\right) + \epsilon \tag{4.88}$$

$$= \max\left(I(\widehat{Y}_3; Y_3 | X_1 X_3 Y_1), I(\widehat{Y}_3; Y_3 | X_2 X_3 Y_2)\right) + \epsilon \tag{4.89}$$

The achievable rate region of the *two-way compress-and-forward strategy* as stated in Theorem 4.3.1 is then given in (4.85) and (4.86). The constraint (4.53) follows by combining (4.68) and (4.89). Note that for (4.89) we make use of Corollary 1 in [107] which says that when the relay is compressing a signal that contains an interference signal which is perfectly known at the decompressing node the interference term can be neglected in the calculations. □

Next, we evaluate the rate region given Theorem 4.3.1 to the special case of additive Gaussian noise channels defined in (4.26)–(4.28).

Proposition 4.3.2 (Two-way compress-and-forward, AWGN). *An achievable rate region of the Gaussian TWRC is given by* $\bigcup \{R_1, R_2\}$ *with*

$$R_1 \leq C\left(P_1|h_0|^2 + \frac{P_1|h_1|^2}{1 + \sigma_c^2}\right) \tag{4.90}$$

$$R_2 \leq C\left(P_2|h_0|^2 + \frac{P_2|h_2|^2}{1 + \sigma_c^2}\right) \tag{4.91}$$

subject to the constraint

$$R_3 \leq \min\left(C\left(\frac{P_3|h_2|^2}{1 + P_1|h_0|^2}\right), C\left(\frac{P_3|h_1|^2}{1 + P_2|h_0|^2}\right)\right). \tag{4.92}$$

The union is over all variances of the compression noise that satisfies $\sigma_c^2 \geq \max\{\sigma_{c1}^2, \sigma_{c2}^2\}$ *with*

$$\sigma_{c1}^2 = \frac{(1 + P_2|h_0|^2)(1 + P_2|h_2|^2) - |P_2 h_0 h_2|^2}{(2^{R_3} - 1)(1 + P_2|h_0|^2)} \tag{4.93}$$

$$\sigma_{c2}^2 = \frac{(1 + P_1|h_0|^2)(1 + P_1|h_1|^2) - |P_1 h_0 h_1|^2}{(2^{R_3} - 1)(1 + P_1|h_0|^2)}. \tag{4.94}$$

where we have chosen all thermal noise variances to be one.

Proof. The probability distribution that maximizes the rates in (4.51) and (4.52) is not known. We therefore choose

$$p(x_1) \sim \mathcal{CN}(0, P_1) \tag{4.95}$$

$$p(x_2) \sim \mathcal{CN}(0, P_2) \tag{4.96}$$

$$p(x_3) \sim \mathcal{CN}(0, P_3) \tag{4.97}$$

$$\tag{4.98}$$

with X_1, X_2 and X_3 statistically independent. The auxiliary random variable $\widehat{Y}_3$ is defined as

$$\widehat{Y}_3 = Y_3 + Z_c = X_1 h_1 + X_2 h_2 + Z_3 + Z_c \tag{4.99}$$

with $Z_c \sim \mathcal{CN}(0, \sigma_c^2)$ and independent of X_1, X_2 and X_3, i.e., σ_c^2 is the variance of the compression (quantization) noise. The achievable rates follow then as

$$R_1 \leq C\left(1 + P_1|h_0|^2 + \frac{P_1|h_1|^2}{1+\sigma_c^2}\right) \tag{4.100}$$

$$R_2 \leq C\left(1 + P_2|h_0|^2 + \frac{P_2|h_2|^2}{1+\sigma_c^2}\right) \tag{4.101}$$

It remains to determine the variance of the compression noise σ_c^2. From (4.89) we have

$$R_3 > I(\widehat{Y}_3; Y_3|X_1X_3Y_1) \tag{4.102}$$

$$= I(\widehat{Y}_3; Y_3|X_1X_3) - I(\widehat{Y}_3; Y_1|X_1X_3) \tag{4.103}$$

$$= h(\widehat{Y}_3; Y_3|X_1X_3Y_1) - h(\widehat{Y}_3; Y_3|X_1X_3Y_3) \tag{4.104}$$

$$= h(Y_3+Z_c, Y_1; Y_3|X_1X_3Y)$$
$$\quad - h(Y_1|X_1X_3) - h(Z_c) \tag{4.105}$$

$$= \log\left((1+P_2|h_2|^2+\sigma_{c1}^2)(1+P_2|h_0|^2) - |P_2h_0h_2|^2\right) \tag{4.106}$$

$$\quad - \log(1+P_2|h_0|^2) - \log(\sigma_{c1}^2). \tag{4.107}$$

Solving (4.107) for σ_{c1}^2 leads to (4.93). The variance σ_{c2}^2 in (4.94) is obtained for R_3 larger than the second term in (4.89):

$$R_3 > I(\widehat{Y}_3; Y_2|X_2X_3) \tag{4.108}$$

$$= \log\left((1+P_1|h_1|^2+\sigma_{c2}^2)(1+P_1|h_0|^2) - |P_1h_0h_1|^2\right) \tag{4.109}$$

$$\quad - \log(1+P_1|h_0|^2) - \log(\sigma_{c2}^2). \tag{4.110}$$

From that one can solve for σ_{c2}^2 in order to obtain (4.94). The rate R_3 follows by the constraint (4.68) which says that both terminals have to decode the index s of the chosen quantization codeword $\widehat{y}_3^n(q,r|s)$ without errors. Hence,

$$R_3 \leq \min\left(C\left(\frac{P_3|h_2|^2}{1+P_1|h_0|^2}\right), C\left(\frac{P_3|h_1|^2}{1+P_2|h_0|^2}\right)\right), \tag{4.111}$$

i.e., terminal T_1 decodes the index s by treating the signal from terminal T_2 as interference and vice versa.

4.4 Amplify-and-forward

We now extend the amplify-and-forward strategy discussed in the previous chapter to the two-way communication scenario. The receive signal at symbol time k at the relay is given as

$$Y_3[k] = h_1 X_1[k] + h_2 X_2[k] + Z_3[k]. \tag{4.112}$$

The receive signal at terminal T_2 is then

$$\begin{aligned}
Y_2[k] &= h_0 X_1[k] + g h_2 Y_3[k-1] + Z_2[k] \\
&= h_0 X_1[k] + g h_1 h_2 X_1[k-1] + Z_2[k] + g h_2 Z_3[k-1]
\end{aligned} \tag{4.113}$$

assuming that terminal T_2 can subtract the contribution of its own transmitted signal $X_2[k-1]$. The receive signal at terminal T_1 follows correspondingly as

$$Y_1[k] = h_0 X_2[k] + g h_1 h_2 X_2[k-1] + Z_1[k] + g h_1 Z_3[k-1]. \tag{4.114}$$

The relay gain g is chosen such that the average transmit power of the relay terminal is P_3, i.e.,

$$g(\alpha) = \sqrt{\frac{P_3}{\alpha P |h_1|^2 + (1-\alpha) P |h_2|^2 + 1}} \tag{4.115}$$

where $P_1 = \alpha P$ and $P_2 = (1-\alpha)P$ and $0 \le \alpha \le 1$. From (4.113) and (4.114) we see that the two-way relay channel with an AF relay decouples into two parallel channels, each being a one-tap inter-symbol interference channel. Therefore, the achievable rate in one direction is given in (3.88).

Proposition 4.4.1 (Two-way amplify-and-forward, AWGN). *An achievable rate region of the Gaussian TWRC is given by the convex hull of the set of all rate pairs* (R_1, R_2) *satisfying*

$$R_1 \le C\left(\frac{c_1 a + \sqrt{(1 + c_1 a)^2 - c_1^2 |b|^2}}{2} \right) \tag{4.116}$$

$$R_2 \le C\left(\frac{c_2 a + \sqrt{(1 + c_2 a)^2 - c_2^2 |b|^2}}{2} \right) \tag{4.117}$$

where

$$a = |h_0|^2 + g(\alpha)^2 |h_1|^2 |h_2|^2 \tag{4.118}$$

$$b = 2g(\alpha)h_1 h_2 \tag{4.119}$$

$$c_1 = \frac{P_1}{1 + g(\alpha)^2 |h_2|^2} \tag{4.120}$$

$$c_2 = \frac{P_2}{1 + g(\alpha)^2 |h_1|^2} \tag{4.121}$$

Furthermore, the rate region defined by (4.116) *and* (4.117) *coincides with the capacity region of the nonrestricted Gaussian TWRC with an amplify-and-forward relay.*

The proof that the rate region is the capacity region of the nonrestricted Gaussian amplify-and-forward TWRC is the same as in [106]. Note that the single-user rates are achieved with $\alpha = 0$ and $\alpha = 1$ and these rates are larger than the rates when both users communicate simultaneously.

4.5 Combined Decode/Compress-and-Forward

In [15], [75] and [16] it was shown that for one-way relay channels decode-and-forward achieves the cut-set bound [74] when the source-relay channel is strong. When the relay-destination channel is strong, then compress-and-forward achieves the cut-set bound. In the two-way relay channel we have the situation that when the relay is near to one of the terminals (see Fig.4.4), employing a pure decode-and-forward strategy or a pure compress-and-forward strategy as given in Propositions 4.2.2 and 4.3.2 results in a low rate for one of the communication directions. For example, when the relay terminal is in the proximity of terminal T_1, using a decode-and-forward strategy gives a large R_1 but a small R_2. Using a compress-and-forward strategy gives a large value for R_2 but a small value for R_1.

Another strategy is to combine both methods: if the relay is near to terminal T_1 it decodes first the message from terminal T_1, treating the signal from terminal T_2 as interference. It subtracts the influence of the decoded message on its received signal, then quantizes and compresses the remaining signal. Terminal T_2 decodes the signal from terminal T_1 treating the bin index sent by the relay terminal as interference, since it cannot determine the bin index from its knowledge of X_2, as we show further below. When the relay is near to terminal T_2 it is the other way around, i.e., compress-and-forward in the forward direction and decode-and-forward in the backward direction. The achievable rate region for this strategy is given in the next proposition directly for AWGN channels.

Proposition 4.5.1 (Two-way decode/compress-and-forward, AWGN). *When* $|h_1| \geq |h_2|$ *we have*

$$\bigcup_{\substack{0 \leq \rho_1 \leq 1 \\ 0 < \gamma < 1}} \{R_1, R_2 : R_1 \leq R_{1,\mathrm{DF}}, R_2 \leq R_{2,\mathrm{CF}}\} \tag{4.122}$$

where

$$R_{1,\mathrm{DF}} = \min \left(C \left(\frac{P_1 |h_1|^2 (1 - |\rho_1|^2)}{1 + P_2 |h_2|^2} \right), \right. \tag{4.123}$$

$$\left. C \left(\frac{P_1 |h_0|^2 + \gamma P_3 |h_2|^2 + 2\rho_{1\mathrm{r}} \sqrt{\gamma P_1 P_3} |h_0| |h_2|}{1 + (1 - \gamma) P_3} \right) \right) \tag{4.124}$$

and

$$R_{2,\mathrm{CF}} = C \left(1 + P_2 |h_0|^2 + \frac{P_2 |h_2|^2}{1 + \sigma_{\mathrm{c}}^2} \right) \tag{4.125}$$

$$\sigma_{\mathrm{c}}^2 \geq \frac{1 + P_2 |h_0|^2 + P_2 |h_2|^2}{(1 - \gamma) P_3 |h_1|^2} \tag{4.126}$$

When $|h_1| < |h_2|$ *we have*

$$\bigcup_{\substack{0 \leq \rho_2 \leq 1 \\ 0 < \gamma < 1}} \{R_1, R_2 : R_1 \leq R_{1,\mathrm{CF}}, R_2 \leq R_{2,\mathrm{DF}}\} \tag{4.127}$$

where

$$R_{1,\mathrm{CF}} = C \left(1 + P_1 |h_0|^2 + \frac{P_1 |h_1|^2}{1 + \sigma_{\mathrm{c}}^2} \right) \tag{4.128}$$

$$\sigma_{\mathrm{c}}^2 \geq \frac{1 + P_1 |h_0|^2 + P_1 |h_1|^2}{(1 - \gamma) P_3 |h_2|^2} \tag{4.129}$$

and

$$R_{2,\mathrm{DF}} = \min \left(C \left(\frac{P_2 |h_2|^2 (1 - |\rho_2|^2)}{1 + P_1 |h_1|^2} \right), \right. \tag{4.130}$$

$$\left. C \left(\frac{P_2 |h_0|^2 + \gamma P_3 |h_1|^2 + 2\rho_{2\mathrm{r}} \sqrt{\gamma P_2 P_3} |h_0| |h_1|}{1 + \gamma P_3} \right) \right) \tag{4.131}$$

Note that in this transmission protocol cancelation of the back-propagating self-interference is not possible for the data stream using compress-and-forward. For the case that the relay is near to terminal T_1 the relay transmits $X_1 + X_3$, i.e., the decoded message of terminal T_1 and the bin index of the compressed signal coming from terminal T_2 (after subtracting the influence of X_1). Terminal T_2 wants to decode X_1 but cannot cancel the interference caused by X_3. The reason is that terminal T_2 knows X_2, but cannot know which bin index was chosen by the relay, as the following calculations show.

After cancelation of the component of terminal T_1 at the relay we need to quantize the signal[7]

$$Y_3' = X_2 + Z_3. \tag{4.132}$$

Thus

$$\widehat{Y}_3 = Y_3' + Z_c = X_2 + Z_3 + Z_c. \tag{4.133}$$

Observe that (X_2, Y_3') are jointly typical (due to random experiment) and $(Y_3', \widehat{Y}_3)$ are jointly typical too (due to quantization strategy). As $X_2 \rightarrow Y_3' \rightarrow \widehat{Y}_3$ is a Markov chain, we have that $(X_2, \widehat{Y}_3)$ are jointly typical too. Thus for a given sequence X_2, the probability that a randomly picked sequence $\widehat{Y}_3$ is jointly typical is given by $2^{-nI(X_2;\widehat{Y}_3)}$. As the quantization codebook contains $2^{n(R_3+R_3')}$ random codewords, we have $2^{n(R_3+R_3')-I(X_2;\widehat{Y}_3)}$ entries in the quantization codebook, which are jointly typical with a given sequence X_2. As we require $R_3+R_3' > I(Y_3'; \widehat{Y}_3)$ for successful quantization and as $I(Y_3'; \widehat{Y}_3) > I(X_2; \widehat{Y}_3)$ in general we have more than one sequence in the quantization codebook that is jointly typical with X_2. Due to random binning the sequences are necessarily in the same bin and terminal T_2 cannot determine the bin index from the knowledge of X_2.

The propositions developed for AWGN channels can be extended to fading channels along the same lines as in Section 3.2.2. The extension of the decode-and-forward scheme is straightforward. For the compress-and-forward scheme the relay needs to have channel knowledge about the direct channel between node one and node two, in order to choose the quantization properly. One interesting point that arises in two-way fading channels is that we do not need a combined decode/compress-and-forward strategy to achieve the cut-set bound when the relay is near to one of the terminals. Since in fading relay channels the CF strategy achieves the cut-set bound when the relay is near by the source or near by the destination, this is also true for fading two-way relay channels. For half-duplex devices we will look at the two-way fading relay channel in the next chapter, where we use this scheme in order to increase the spectral efficiency of half duplex relaying.

---

[7]We assume for all channels $h_0 = h_1 = h_2 = 1$.

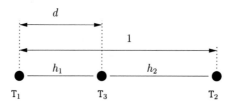

Fig. 4.4: Simplified relay network

4.6 Numerical Examples

We compare the different strategies with respect to sum-rate for the simple network geometry depicted in Fig.4.4. We assume the channel gains to be $h_0 = 1$, $h_1 = 1/d^{\alpha/2}$, $h_2 = 1/(1-d)^{\alpha/2}$, where d is the distance between terminal T_1 and relay terminal T_3 and $\alpha = 2$ is the path loss exponent. We choose $P_1 = P_2 = P_3 = 10$ and zero mean, unit variance Gaussian noise variables. An upper bound on the maximum sum-rate is obtained by applying the cut-set bound [74] to the two-way relay channel and is given as

$$R_{\text{sum}} < \max_{0 \leq \gamma \leq 1} \Big(\max_{0 \leq \rho_{1r} \leq 1} \min \big(C \left(P_1(|h_1|^2 + |h_0|^2) \left(1 - \rho_{1r}^2 \right) \right),$$
$$C \left(P_1|h_0|^2 + \gamma P_3|h_2|^2 + 2\rho_{1r}\sqrt{\gamma P_1 P_3}|h_0||h_2| \right) \big)$$
$$+ \max_{0 \leq \rho_{2r} \leq 1} \min \big(C \left(P_2(|h_2|^2 + |h_0|^2) \left(1 - \rho_{2r}^2 \right) \right),$$
$$C \left(P_2|h_0|^2 + \overline{\gamma} P_3|h_1|^2 + 2\rho_{2r}\sqrt{\overline{\gamma} P_2 P_3}|h_0||h_1| \right) \big) \Big). \qquad (4.134)$$

In Fig.4.5 we compare the sum-rate of the different strategies with the cut-set upper bound of the TWRC [108]. We see that the combined strategy of Proposition 4.5.1 achieves the cut-set bound, when the relay is in the proximity of terminal T_1 or T_2 but has loose performance when the relay is somewhere in the middle. We have also plotted an upper for the combined DF/CF strategy which assumes that the terminals T_1 and T_2 can cancel the bin index perfectly. In this case the interference terms in the denominator of (4.124) and (4.131) are zero. In the middle region the best performance is achieved by the compress-and-forward strategy of Proposition 4.3.2. We see that the amplify-and-forward strategy also performs well in this region. The reason is, that in both strategies the relay has not to decode the messages from terminal T_1 and T_2 and is therefore not limited by the multiple access sum-rate. The decode-and-forward strategy based on block Markov superposition coding performs better than ordinary decode-and-forward (without

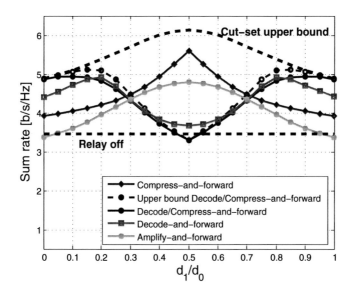

Fig. 4.5: Sum-rates of AWGN two-way relay channel for different relaying strategies

coherent cooperation of relay and terminals) only, when the relay is not too far away from one of the terminals. In the middle region both schemes coincide because the optimal correlation coefficients become zero. For comparison, we also plotted the sum-rate, when the relay does not help in the communication process. We see that in most cases the cooperative strategies outperform the noncooperative strategy. For all curves the parameters ρ_1, ρ_2, γ were optimized numerically.

5 Spectral Efficient Protocols for Two-hop Half-Duplex Relay Channels

As discussed in Chapter 1 the analysis and design of cooperative transmission protocols for wireless networks has recently attracted a lot of interest. Of particular interest are two-hop channels where a relay terminal assists in the communication between a source terminal and a destination terminal. We discussed several relaying strategies in Chapters 3 and 4. In certain cases, there is no direct connection between source terminal and destination terminal, for example due to shadowing or too large separation between source and destination. In such cases the relay has always to be used in order to transmit data from the source to the destination.

For example, in [109] the authors consider a relay network with one source and one destination both equipped with M antennas and K half-duplex relays each equipped with $N \geq 1$ antennas. In the absence of a direct link between source and destination and the use of amplify-and-forward (AF) relays the authors show that the capacity scales as $\frac{M}{2} \log(\text{SNR})$ for high signal-to-noise ratios (SNR) when the number of relays K grows to infinity. The pre-log factor $\frac{1}{2}$ is induced by the half-duplex signaling (two channel uses) and causes a substantial loss in spectral efficiency. Further half-duplex relaying protocols with a pre-log factor $\frac{1}{2}$ can be found in [110] and the references therein. One way to avoid the pre-log factor $\frac{1}{2}$ is to use a full-duplex relay that may receive and transmit at the same time and frequency [14], but such a relay is difficult to implement. Large differences in the signal power of the transmitted and the received signal drive the relay's analog amplifiers in the receive chain into saturation and cause problems to the cancelation of the self-interference. In [111] the authors address the half-duplex loss by proposing a spatial reuse of the relay slot. They consider a base station that transmits K messages to K users and their corresponding relays in K orthogonal time slots. In time slot $K + 1$ all relays retransmit their received signal, causing interference to the other users. The capacity of a single connection (base station to user) has then a pre-log factor $\frac{K}{K+1}$ instead of $\frac{1}{2}$. A similar scheme was proposed in [112] where the authors study the range extension potential of fixed half-duplex decode-and-forward relays in a cellular network. In order to prevent a loss in spectral efficiency (throughput) due to the half-duplex constraint of the relays the authors propose to reuse existing channels in the cellular network for the relay transmissions, thereby causing co-channel interference. They propose a relay channel selection scheme that keeps the level of co-channel interference low and show that significant throughput improvements can be achieved. Another solution is presented in [113] where the authors propose a transmission scheme with two half-duplex AF relays that alternately forward messages from a source to a destination. In order to decrease the inter-relay interference, one relay performs interference cancelation. This cooperation scheme turns the equivalent channel between source and destination into a frequency-selective channel. A maximum likelihood sequence estimator at the destination is applied to extract the introduced diversity, an idea which is known as delay diversity [114]. However, the authors did not study the achievable rate of this scheme. A similar protocol with two alternating relays was introduced in [115] where the transmitting terminals, i.e., source, relay one and relay two, use orthogonal direct sequence spreading codes. By that the destination is able to separate the signals from the source and both relays. Also a relay may separate the signals transmitted by the source and the other relay. It was shown that with this system full diversity order of two is achievable without sacrificing bandwidth (besides the spreading). However, the system utilizes three different codes (code multiplex) to avoid interference between the transmissions. During writing this thesis we discovered the work in [116] which investigates a similar problem as we do in our work. The authors consider two relays that alternately receive and transmit data and where the direct channel is also used for transmissions. The relays operate as half-duplex decode-and-forward transceivers with interference cancelation and the destination employs a V-BLAST receiver to resolve signal collisions between relay and source signals.

In this chapter we propose two half-duplex relaying protocols that mitigate the loss in spectral efficiency due to the half-duplex operation of the relays. *Firstly*, we propose a relaying protocol where a synchronous bidirectional connection (two-way protocol) between two terminals (e.g., two wireless routers) is established using one half-duplex AF or DF relay. Hereby, the achievable rate in one direction suffers still from the pre-log factor $\frac{1}{2}$ but since two connections are realized in the same physical channel we can achieve a sum-rate that is above the single user rate for the half-duplex relay channel. We extend the two-way protocol to a multi-user scenario, where multiple terminals communicate with multiple partner terminals via several orthogonalize-and-forward (OF) relays. *Secondly*, we consider a similar relaying scheme as in [113] and [115] but with the difference that the source and the relay operate in the same physical channel without using orthogonal spreading codes and that our AF relays only amplify-and-forward their received signals (no cancelation of the inter-relay interference at one of the relays as in [113]). We propose to employ successive decoding at the destination with partial or full cancelation of the inter-relay interference. We also analyze the achievable rate when DF relays are used. It is shown that this protocol can recover a significant portion of the half-duplex loss (pre-log factor $\frac{1}{2}$) for both AF and DF relays.

5.1 Spectral Efficiency of Two-hop Half-Duplex Relaying Protocols

In this section we first review three half-duplex protocols used in wireless relaying. We consider the case where one source terminal communicates with one destination terminal with the help of one relay terminal but (in contrast to Chapter 3) there is no direct connection between source and destination, for example due to shadowing or too large separation [109], [61]. In that case the source cannot transmit in the second time slot (relay-transmit period), since this data would be lost. The need for two time slots to transmit one unit of information leads to an unavoidable loss in spectral efficiency due to the pre-log factor one-half in the corresponding capacity expression. We look at the spectral efficiency that is achievable when the relay uses an *amplify-and-forward* (AF) strategy or a *decode-and-forward* (DF) strategy. We assume that all terminals operate in half-duplex mode [77], [16]. We also consider a network of terminals where a number of sources communicates with a number of destinations with the help of a certain number of relays. In this case wee look at the spectral efficiency of a scheme that was introduced in [91] [117], and which we call here *orthogonalize-and-forward* (OF). Again, the assumptions that there is no direct connection between source terminals and destination terminals and the half-duplex constraint lead to a factor one-half loss in spectral efficiency.

5.1.1 Amplify-and-forward

A source terminal T_1 transmits in the first time slot its information symbols to a relay terminal T_3. The relay amplifies the received symbol (including noise) according to its available average transmit power and forwards its in the second time slot to the destination terminal T_2. The random variables of the channel are:

- x_1: transmit symbol of the source that is i.i.d $\mathcal{CN}(0, P_1)$ (Gaussian code book),

- y_2, y_3: complex received symbol at the destination and relay, respectively,

- g: real scaling coefficient of the relay,

- h_1, h_2: complex channel gain between source and relay (first hop) and between relay and destination (second hop), respectively,

- n_2, n_3: complex, zero-mean, additive white Gaussian noise at the destination and relay, with variances σ_2^2 and σ_3^2, respectively.

Notice that all random variables are independent of each other. The input-output relation after time slot k is given by[1]

$$y_2[k] = n_2[k] \tag{5.1}$$

$$y_3[k] = h_1[k]x_1[k] + n_3[k]. \tag{5.2}$$

The relay observes $y_3[k]$ and scales it by[2]

$$g[k] = \sqrt{\frac{P_3}{P_1|h_1[k]|^2 + \sigma_3^3}} \tag{5.3}$$

where P_1 is the average transmit power of the source and P_3 the average transmit power of the relay [27]. The input-output relation after the next time slot $k + 1$ is then

$$y_2[k+1] = h_2[k+1]g[k]h_1[k]x_1[k] + h_2[k+1]g[k]n_3[k] + n_2[k+1] \tag{5.4}$$

$$y_3[k+1] = 0. \tag{5.5}$$

The ergodic information rate of this scheme is given by

$$R_{\mathrm{AF}} = \frac{1}{2}\mathcal{E}\left\{\log\left(1 + \frac{P_1|h_2gh_1|^2}{\sigma_2^2 + \sigma_3^3|h_2g|^2}\right)\right\}. \tag{5.6}$$

The pre-log factor one-half follows because the relay operates in half-duplex mode and two time slots are needed to transmit the information from the source to the destination.

5.1.2 Decode-and-forward

The relay decodes the message sent by the source, re-encodes it by using the same or a different codebook and forwards the message to the destination. The input-output relation after time slot

---

[1]The duration of one time slot is normalized to $T = 1$.

[2]Different possibilities for the choice of the relay gain are known from literature, depending on the amount of channel knowledge at the relay.

k is given by

$$y_2[k] = n_2[k] \tag{5.7}$$

$$y_3[k] = h_1[k]x_1[k] + n_3[k] \tag{5.8}$$

and after time slot $k + 1$

$$y_2[k + 1] = h_2[k + 1]x_3[k + 1] + n_2[k + 1] \tag{5.9}$$

$$y_3[k + 1] = 0 \tag{5.10}$$

where $x_3 \sim \mathcal{CN}(0, P_3)$ is the transmit symbol of the relay. The ergodic information rate of this scheme is given by

$$R_{\mathrm{DF}} = \frac{1}{2} \min \left\{ \mathcal{E} \left\{ \log \left(1 + \frac{P_1|h_1|^2}{\sigma_3^2} \right) \right\}, \mathcal{E} \left\{ \log \left(1 + \frac{P_3|h_2|^2}{\sigma_2^2} \right) \right\} \right\}. \tag{5.11}$$

Note that since the direct connection between source and destination is not available it is not possible to use a decode-and-forward scheme based on superposition coding as described in Chapter 3 and [14], [16], where the signals from the relay and the source coherently add up at the destination. Furthermore, the rate given in (5.11) is exactly the ergodic capacity of the half-duplex relay channel with no direct connection, which can be easily seen by applying the cut-set upper bound [14] and by inspecting that (5.11) is equal to this upper bound.

5.1.3 Orthogonalize-and-forward

Consider now a network with $2N + K$ terminals, see Fig.5.1. The network is divided into three sets: N terminals in $\mathcal{T}_1$ want to transmit messages to N terminals in $\mathcal{T}_2$ with the help of K relay terminals in $\mathcal{T}_3$. The random variables of the channel are:

- $\mathbf{x}_1$: $N \times 1$ dimensional vector that contains the transmit symbols of all source terminals in $\mathcal{T}_1$. The elements of $\mathbf{x}_1$ are i.i.d. $\mathcal{CN}(0, P_1)$,

- $\mathbf{y}_2$: $N \times 1$ dimensional vector of received symbols at the destination terminals in $\mathcal{T}_2$,

- $\mathbf{y}_3$: $K \times 1$ dimensional vector of received symbols at the relay terminals in $\mathcal{T}_3$,

- $\mathbf{H}_1$: $K \times N$ dimensional channel matrix between terminals in $\mathcal{T}_1$ and $\mathcal{T}_3$,

- $\mathbf{H}_2$ $N \times K$ dimensional channel matrix between terminals in $\mathcal{T}_3$ and $\mathcal{T}_2$,

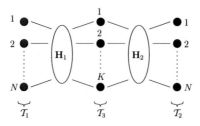

Fig. 5.1: Muli-user relaying with distributed zero-forcing relays (orthogonalize-and-forward)

- $\mathbf{n}_2$: $N \times 1$ dimensional vector of i.i.d. zero-mean, additive white Gaussian noise terms each with variance σ_2^2 at the destination terminals in $\mathcal{T}_2$,

- $\mathbf{n}_3$: $K \times 1$ dimensional vector of i.i.d. zero-mean, additive white Gaussian noise terms each with variance σ_3^2 at the relay terminals in $\mathcal{T}_2$.

Again all random variables are independent of each other. The input-output relation after time slot k is

$$\mathbf{y}_2[k] = \mathbf{n}_2[k] \tag{5.12}$$

$$\mathbf{y}_3[k] = \mathbf{H}_1[k]\mathbf{x}_1[k] + \mathbf{n}_3[k]. \tag{5.13}$$

Each relay $\mathrm{T}_{3i} \in \mathcal{T}_3$, $i = 1, 2, \ldots, K$ scales its observation (which is a superposition of the transmit signals of all sources $\mathrm{T}_{1i} \in \mathcal{T}_1$, $i = 1, 2, \ldots, N$) by a coefficient $g_i[k]$ such that an average sum power constraint among the relays is fulfilled and such that the overall channel between the sources and the destinations becomes diagonal. The input-output relation after time slot $k + 1$ is

$$\mathbf{y}_2[k+1] = \mathbf{H}_2[k+1]\mathbf{G}[k]\mathbf{H}_1[k]\mathbf{x}_1[k] + \mathbf{H}_2[k+1]\mathbf{G}[k]\mathbf{n}_3[k] + \mathbf{n}_2[k+1] \tag{5.14}$$

$$\mathbf{y}_3[k+1] = 0. \tag{5.15}$$

We choose the diagonal gain matrix $\mathbf{G}[k] = \mathrm{diag}\,(g_1[k], g_2[k], \ldots, g_K[k])$ such that the transmissions between nodes in $\mathcal{T}_1$ and $\mathcal{T}_2$ become interference-free. For this purpose we define the interference matrix $\mathbf{H}_{\text{int}}$ with dimensions $N(N-1) \times K$ to be[3]

$$\mathbf{H}_{\text{int}} = \left[\mathbf{h}_1^{(1)} \odot \mathbf{h}_2^{(2)} , \ldots , \mathbf{h}_1^{(1)} \odot \mathbf{h}_q^{(2)} , \mathbf{h}_2^{(1)} \odot \mathbf{h}_1^{(2)} , \mathbf{h}_2^{(1)} \odot \mathbf{h}_3^{(2)} , \ldots , \mathbf{h}_2^{(1)} \odot \mathbf{h}_q^{(2)} , \ldots \right]^{\text{T}}$$

$$(5.16)$$

i.e., the columns of $\mathbf{H}_{\text{int}}$ are defined by

$$\mathbf{h}_p^{(1)} \odot \mathbf{h}_q^{(2)} \tag{5.17}$$

for all $p, q \in \{1, 2, \ldots, N\}$ and $p \neq q$, where $\mathbf{h}_p^{(1)}$ is the vector containing the channel gains from the pth node in $\mathcal{T}_1$ to all relays in $\mathcal{T}_3$ (i.e., the pth column of $\mathbf{H}_1$) and $\mathbf{h}_q^{(2)}$ is the vector containing the channel gains from the qth node in $\mathcal{T}_2$ to all relays in $\mathcal{T}_3$ (i.e., the qth column of $\mathbf{H}_2^{\text{T}}$). The two-hop channel $\mathbf{H}_2\mathbf{G}\mathbf{H}_1$ becomes diagonal if $\mathbf{g} = \text{diag}(\mathbf{G}) = (g_1, g_2, \ldots, g_K)^{\text{T}}$ lies in the null space of the interference matrix $\mathbf{H}_{\text{int}}$ (*zero-forcing*). Let $r = \text{rk}(\mathbf{H}_{\text{int}}) = \min\{N^2 - N, K\}$ be the rank of the matrix $\mathbf{H}_{\text{int}}$ and define the singular value decomposition

$$\mathbf{H}_{\text{int}} = \mathbf{U}\mathbf{D} \left[\mathbf{V}^{(r)} \ \mathbf{V}^{(0)} \right]^{\text{H}} \tag{5.18}$$

where $\mathbf{V}^{(r)}$ contains the first r right singular vectors of $\mathbf{H}_{\text{int}}$ and $\mathbf{V}^{(0)}$ the last $K - r$ right singular vectors. The columns of $\mathbf{V}^{(0)}$ form an orthonormal basis for the null space of $\mathbf{H}_{\text{int}}$, i.e., $\mathbf{V}^{(0)} = \text{null}(\mathbf{H}_{\text{int}})$. A sufficient condition for the null space to be non-empty is

$$K \geq N(N-1) + 1 \tag{5.19}$$

and we refer to it as *minimum relay configuration*. The orthogonalize-and-forward gain vector $\mathbf{g}$ is obtained by projecting any gain vector $\widetilde{\mathbf{g}}$ onto the null space of $\mathbf{H}_{\text{int}}$. In [91] it was shown that for $K \to \infty$ the choice $\widetilde{\mathbf{g}} = \text{diag}\left(\mathbf{H}_2\mathbf{H}_1\right)$ diagonalizes the two-hop channel $\mathbf{H}_2\mathbf{G}\mathbf{H}_1$. For finite number of relays $K < \infty$ we choose

$$\mathbf{g} = c\mathbf{Z}\mathbf{Z}^{\text{H}}\widetilde{\mathbf{g}} \tag{5.20}$$

where $\mathbf{Z} = \mathbf{V}^{(0)}$ and c is chosen such that the average sum power constraint of the relays is met. In order to compute $\mathbf{g}$ each relay has to know all channel gains of the network. In [118] it is shown how this channel information may be distributed when the relays have access to a powerline distribution system. Using $\mathbf{g}$ the product channel $\mathbf{H}_{12} = \mathbf{H}_2\mathbf{G}\mathbf{H}_1$ with $\mathbf{G} = \text{diag}(\mathbf{g})$ becomes diagonal and the transmissions between terminals in $\mathcal{T}_1$ and $\mathcal{T}_2$ interference-free. However, the

---

[3]For ease of notation we drop the time index k.

noise terms at the destination nodes in T_2 are still correlated. For separate (noncooperating) decoding at the terminals in T_1 the network's sum-rate follows as

$$R_{\text{sum,sep}}^{\text{OF}} = \frac{1}{2} \mathcal{E} \left\{ \log \det \left(\mathbf{I}_N + P_1 \mathbf{D}_2^{-1} \mathbf{H}_{12} \mathbf{H}_{12}^{\text{H}} \right) \right\}. \tag{5.21}$$

The factor one-half in the pre-log is again due to the use of two time slots. The diagonal matrix $\mathbf{D}_2$ contains the diagonal elements of the noise covariance matrix

$$\mathbf{D}_2 = \text{diag} \left(\mathcal{E} \left\{ \mathbf{n}_2 \mathbf{n}_2^{\text{H}} \right\} \right) = \text{diag} \left(\mathbf{R}_2 \right) = \text{diag} \left(\sigma_3^2 \mathbf{H}_2 \mathbf{G} \mathbf{G}^{\text{H}} \mathbf{H}_2^{\text{H}} + \sigma_2^2 \mathbf{I}_N \right). \tag{5.22}$$

Since the noise samples at terminals T_2 are correlated we look also at the case of cooperative terminals in T_2 (i.e., joint decoding). The achievable sum-rate is then given by

$$R_{\text{sum,joint}}^{\text{OF}} = \frac{1}{2} \mathcal{E} \left\{ \log \det \left(\mathbf{I}_N + P_1 \mathbf{R}_2^{-1} \mathbf{H}_{12} \mathbf{H}_{12}^{\text{H}} \right) \right\}. \tag{5.23}$$

Clearly, $R_{\text{sum,joint}}^{\text{ZF}} \geq R_{\text{sum,sep}}^{\text{ZF}}$, since the noise samples are correlated at the different users, but the rate loss due to separate decoding is small as the numerical examples in Section 5.4 will show.

5.2 Increasing the Spectral Efficiency with Two-way Relaying

The relaying protocols discussed in the previous section suffer from a loss in spectral efficiency due to the half-duplex constraint of the terminals. A possibility to increase the spectral efficiency of such a relay network is to consider a bidirectional (two-way) communication between two terminals whereas the relay assists in the two-way communication. An example for such a scenario may be the communication between two wireless routers that communicate with each other through the help of a relay terminal due to the lack of a direct connection[4]. The two-way communication problem was first studied by Shannon in [105], which is considered as the first study of a network information theory problem. In the previous chapter we studied the achievable rate regions for the general full-duplex two-way relay channel (including direct link). In this section we show how the half-duplex relaying strategies discussed in the previous section can be extended to the two-way case and analyze the achievable sum-rates.

---

[4]Note that a direct channel could not be used in a two-way protocol with half-duplex terminals, since both terminals transmit simultaneously and also receive simultaneously.

5.2.1 Amplify-and-forward

Terminal T_1 wants to transmit messages to terminal T_2 and vice versa. Since there is no direct connection between the terminals all traffic goes trough relay terminal T_3. The proposed relaying scheme works as follows: in time slot k both terminals T_1 and T_2 transmit their symbols to relay T_3. The relay receives in time slot k

$$y_3[k] = h_1[k]x_1[k] + h_2[k]x_2[k] + n_3[k] \tag{5.24}$$

where $x_2 \sim \mathcal{CN}(0, P_2)$ is the transmit symbol of terminal T_2. The relay scales the received signal by

$$g[k] = \sqrt{\frac{P_3}{P_1|h_1|^2 + P_2|h_2|^2 + \sigma_3^2}} \tag{5.25}$$

in order to meet its average transmit power constraint. It then broadcasts the signal in the next time slot to both destinations. The input-output relation for the $T_1 \rightarrow T_3 \rightarrow T_2$ communication direction is

$$y_2[k+1] = h_2[k+1]g[k]h_1[k]x_1[k] + h_2[k+1]g[k]h_2[k]x_2[k] + h_2[k+1]g[k]n_3[k] + n_2[k+1] \tag{5.26}$$

and for the $T_1 \leftarrow T_3 \leftarrow T_2$ direction

$$y_1[k+1] = h_1[k+1]g[k]h_2[k]x_2[k] + h_1[k+1]g[k]h_1[k]x_1[k] + h_1[k+1]g[k]n_3[k] + n_1[k+1]. \tag{5.27}$$

Note that we assume reciprocity for the channels between the terminals. Since nodes T_1 and T_2 know their own transmitted symbols they can subtract the *back-propagating self-interference* in (5.26) and (5.27) prior to decoding, assuming perfect knowledge of the corresponding channel coefficients. The sum-rate of this protocol is then given by

$$R_{\text{sum}}^{\text{AF}} = \frac{1}{2}\mathcal{E}\left\{\log\left(1 + \frac{P_1|h_2gh_1|^2}{\sigma_2^2 + \sigma_3^2|h_2g|^2}\right)\right\} + \frac{1}{2}\mathcal{E}\left\{\log\left(1 + \frac{P_2|h_2gh_1|^2}{\sigma_1^2 + \sigma_3^2|h_1g|^2}\right)\right\}. \tag{5.28}$$

The transmission in each direction suffers still from the pre-log factor $\frac{1}{2}$. However, the half-duplex constraint can here be exploited to establish a bidirectional connection between two nodes and to increase the sum-rate of the network.

5.2.2 Decode-and-forward

We consider now two-way communication between terminals T_1 and T_2 via a half-duplex DF relay T_3. In time slot k both terminals T_1 and T_2 transmit their symbols to relay T_3. The relay receives in time slot k

$$y_3[k] = h_1[k]x_1[k] + h_2[k]x_2[k] + n_3[k] \tag{5.29}$$

and decodes the symbols $x_1[k]$ and $x_2[k]$. In time slot $k + 1$ the input-output relation for the $T_1 \rightarrow T_3 \rightarrow T_2$ communication direction is

$$y_2[k] = h_2[k + 1]x_3[k + 1] + n_2[k] \tag{5.30}$$

and for the $T_1 \leftarrow T_3 \leftarrow T_2$ direction

$$y_1[k] = h_1[k + 1]x_3[k + 1] + n_1[k] \tag{5.31}$$

where

$$x_3[k + 1] = \sqrt{\beta}\,\widehat{x}_1[k] + \sqrt{1 - \beta}\,\widehat{x}_2[k] \tag{5.32}$$

is the superposition of the decoded symbols from the previous time slot. The relay may use an average transmit power of βP_3 for the forward direction and $(1 - \beta)P_3$ for the backward direction. Since terminal T_1 knows $x_1[k]$ and terminal T_2 knows $x_2[k]$ these symbols (back-propagating self-interference) can be subtracted at the respective terminals prior to decoding of the symbol transmitted by the partner terminal. Again we assume channel reciprocity and that the relay can decode x_1 and x_2 without errors, i.e, $\widehat{x}_1[k] = x_1[k]$ and $\widehat{x}_2[k] = x_2[k]$. The sum-rate of this protocol is then given by

$$R_{\text{sum}}^{\text{DF}} = \max_\beta \min\left(R_{\text{MA}}, R_1(\beta) + R_2(1 - \beta)\right) \tag{5.33}$$

where

$$R_{\text{MA}} = \frac{1}{2}C\left(P_1|h_1|^2 + P_2|h_2|^2\right) \tag{5.34}$$

$$R_1(\beta) = \frac{1}{2}\min\left(C\left(P_1|h_1|^2\right), C\left(\beta P_3|h_2|^2\right)\right) \tag{5.35}$$

$$R_2(1 - \beta) = \frac{1}{2}\min\left(C\left(P_2|h_2|^2\right), C\left((1 - \beta)P_3|h_1|^2\right)\right) \tag{5.36}$$

and where $C(x) = \mathcal{E}\{\log(1 + x)\}$. If the relay does not have any knowledge of the channels[5] to terminal T_1 and T_2 it chooses $\beta = \frac{1}{2}$. For the case the relay has some channel knowledge about h_1 and h_2 (it may learn it during the previous transmission from the terminals to the relay) β may be chosen such, that the sum-rate is maximized. Clearly, the power allocation that maximizes the sum-rate depends on the degree of channel knowledge (instantaneous channel gains, second order statistics, etc). For simplicity we choose $P_1 = P_2 = P_3 = P$ and assume that the relay has perfect knowledge of the backward and forward channels h_1 and h_2. The optimal power allocation is then found to be:

$$
\beta^* = \begin{cases}
\min\left\{\frac{|h_1|^2}{|h_2|^2}, \frac{1}{2P|h_1|^2} - \frac{1}{2P|h_2|^2} + \frac{1}{2}\right\}, & |h_1|^2 \leq |h_2|^2; \\
\max\left\{1 - \frac{|h_2|^2}{|h_1|^2}, \frac{1}{2P|h_1|^2} - \frac{1}{2P|h_2|^2} + \frac{1}{2}\right\}, & |h_1|^2 > |h_2|^2
\end{cases}
\tag{5.37}
$$

Proof. See Appendix 8.

Assume now that the relay has only knowledge of the channel distributions $p_{H_1}(h_1)$ and $p_{H_2}(h_2)$. The optimal power allocation is then given by:

$$
\beta^* = \begin{cases}
\min\left\{\frac{\mathcal{E}\{|h_1|^2\}}{\mathcal{E}\{|h_2|^2\}}, \frac{1}{2P\mathcal{E}\{|h_1|^2\}} - \frac{1}{2P\mathcal{E}\{|h_2|^2\}} + \frac{1}{2}\right\}, & C\left(P|h_1|^2\right) \leq C\left(P|h_2|^2\right); \\
\max\left\{1 - \frac{\mathcal{E}\{|h_2|^2\}}{\mathcal{E}\{|h_1|^2\}}, \frac{1}{2P\mathcal{E}\{|h_1|^2\}} - \frac{1}{2P\mathcal{E}\{|h_2|^2\}} + \frac{1}{2}\right\}, & C\left(P|h_1|^2\right) > C\left(P|h_2|^2\right)
\end{cases}
\tag{5.38}
$$

Proof. The result follows immediately from the proof in Appendix 8 by replacing the mutual information I with ergodic mutual information, i.e., $\mathcal{E}\{I\}$.

Note that the power allocation (5.38) has the same structure as in (5.37), with the difference that only second-order statistics is necessary for the power allocation. In Fig.5.2 we plot the optimum power allocation for a linear one-dimensional relay network, as illustrated in Fig.5.8. For simplicity we assume only channels with path loss but no small-scale fading. We plot the optimum power allocation for transmit powers $P = 1$ and $P = 0.1$. As the relay moves towards terminal T_1, i.e., $d \to 0$, more relay power is spent for the $T_1 \to T_3 \to T_2$ transmission and less power for the reverse $T_1 \leftarrow T_3 \leftarrow T_2$ transmission. The reason is that the link capacity from terminal T_2 to relay T_3 is small and dominates the overall capacity for the $T_1 \leftarrow T_3 \leftarrow T_2$ transmission, irrespective of the relay power allocated to that transmission. But at the point where the mutual information of the second hop channel $T_3 - -T_2$ in the forward direction becomes larger than the mutual information of the first hop channel $T_1 - -T_3$, the relay stops to allocate

---

[5]For example, when the channels change i.i.d. from time slot to time slot, the channel knowledge learned during the multiple access phase cannot not be used for the broadcast phase.

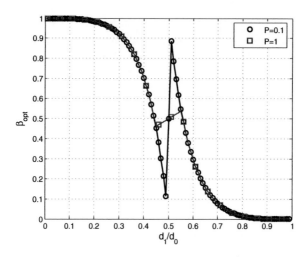

Fig. 5.2: Optimum power allocation in two-way decode-and-forward relaying

more power into the second hop $T_3 - -T_2$ of the forward transmission and starts to increase the power allocated to the second hop $T_3 - -T_1$ of the reverse direction. The reversal of the power allocation is more dominant for low transmit powers (or equivalently, for low SNR). When the relay is half-way between both terminals the relay allocates equal power to both communication directions.

5.2.3 Orthogonalize-and-forward

We apply the two-way communication scheme to the network with N two-hop communication links and K relays. Like in Section 5.1.3 we divide the network into three sets of nodes: nodes in T_1 want to transmit messages to nodes in T_2 and vice versa and the set T_3 contains the relay nodes. The input-output relation for the $T_1 \rightarrow T_3 \rightarrow T_2$ communication is

$$
\begin{aligned}
\mathbf{y}_2[k+1] = &\mathbf{H}_2[k+1]\mathbf{G}[k]\mathbf{H}_1[k]\mathbf{x}_1[k] + \mathbf{H}_2[k+1]\mathbf{G}[k]\mathbf{H}_2^{\mathrm{T}}[k]\mathbf{x}_2[k] \\
&+\mathbf{H}_2[k+1]\mathbf{G}[k]\mathbf{n}_3[k] + \mathbf{n}_2[k+1]
\end{aligned}
\tag{5.39}
$$

and for the $T_1 \leftarrow T_3 \leftarrow T_2$ communication

$$\mathbf{y}_1[k+1] = \mathbf{H}_1^{\mathrm{T}}[k+1]\mathbf{G}[k]\mathbf{H}_2^{\mathrm{T}}[k]\mathbf{x}_2[k] + \mathbf{H}_1^{\mathrm{T}}[k+1]\mathbf{G}[k]\mathbf{H}_1[k]\mathbf{x}_1[k]$$
$$+\mathbf{H}_1^{\mathrm{T}}[k+1]\mathbf{G}[k]\mathbf{n}_3[k] + \mathbf{n}_1[k+1]. \tag{5.40}$$

The $N \times 1$ vectors $\mathbf{x}_1$ and $\mathbf{x}_2$ comprise the transmitted symbols from nodes in T_1 and T_2, respectively. Each element (symbol) of $\mathbf{x}_1$ and $\mathbf{x}_2$ is taken from a Gaussian codebook with average transmit power P_1 and P_2, respectively. The $N \times 1$ vectors $\mathbf{y}_1$ and $\mathbf{y}_2$ comprise the received symbols of nodes in T_1 and T_2 in the second time slot. In the following we discuss how to choose the diagonal gain matrix $\mathbf{G}$ such that the transmissions between nodes in T_1 and T_2 become interference-free in both communication directions. Comparing (5.39) with (5.14), we see that a receiving terminal in T_2 suffers additionally from back-propagating interference caused by its neighbor terminals located in the same set. Each terminal knows only its own symbol (which was transmitted by this terminal in time slot k) and can subtract this contribution from the received signal in time slot $k+1$. The symbols transmitted by the neighbor terminals are unknown and cannot be subtracted (since the terminals in each set do not cooperate and therefore have no knowledge about the other's transmitted symbols). The gain matrix $\mathbf{G}$ has therefore to be chosen such that the channels $\mathbf{H}_2\mathbf{G}\mathbf{H}_2^{\mathrm{T}}$ and $\mathbf{H}_1^{\mathrm{T}}\mathbf{G}\mathbf{H}_1$ become diagonal too. For this purpose the interference matrix (5.16) has to be extended to a $K \times 2N(N-1)$ dimensional matrix

$$\mathbf{H}_{\mathrm{int,bi}} = [\mathbf{H}_{\mathrm{int}}, \mathbf{H}_{\mathrm{int,1}}, \mathbf{H}_{\mathrm{int,2}}] \tag{5.41}$$

where

$$\mathbf{H}_{\mathrm{int,1}} = \left[\mathbf{h}_1^{(1)} \odot \mathbf{h}_2^{(1)}, \ldots, \mathbf{h}_1^{(1)} \odot \mathbf{h}_N^{(1)}, \mathbf{h}_2^{(1)} \odot \mathbf{h}_3^{(1)}, \ldots, \mathbf{h}_2^{(1)} \odot \mathbf{h}_N^{(1)}, \ldots, \mathbf{h}_{N-1}^{(1)} \odot \mathbf{h}_N^{(1)}\right]$$
$$\mathbf{H}_{\mathrm{int,2}} = \left[\mathbf{h}_1^{(2)} \odot \mathbf{h}_2^{(2)}, \ldots, \mathbf{h}_1^{(2)} \odot \mathbf{h}_N^{(2)}, \mathbf{h}_2^{(2)} \odot \mathbf{h}_3^{(2)}, \ldots, \mathbf{h}_2^{(2)} \odot \mathbf{h}_N^{(2)}, \ldots, \mathbf{h}_{N-1}^{(2)} \odot \mathbf{h}_N^{(2)}\right].$$

are both $K \times \frac{N(N-1)}{2}$ dimensional matrices. Note that for example in $\mathbf{H}_{\mathrm{int,1}}$ it suffices to consider only $\mathbf{h}_1^{(1)} \odot \mathbf{h}_2^{(1)}$, whereas $\mathbf{h}_2^{(1)} \odot \mathbf{h}_1^{(1)}$ does not appear in the definition of the interference matrix. The reason is that the interference channel from terminal 1 to terminal 2 is the same as the interference channel from terminal 2 to terminal 1 when we assume channel reciprocity. This symmetry leads to a $K \times 2N(N-1)$ dimensional interference matrix rather than to a $K \times 3N(N-1)$ interference matrix. Therefore, for a non-empty nullspace the minimum relay configuration becomes

$$K \geq 2N(N-1) + 1 \tag{5.42}$$

which is a sufficient but not necessary condition for a nonempty null space. The network's two-way sum-rate follows as

$$R_{\text{sum,sep}}^{\text{OF}} = \frac{1}{2}\mathcal{E}\left\{\log\det\left(\mathbf{I}_N + P_1\mathbf{D}_2^{-1}\mathbf{H}_{12}\mathbf{H}_{12}^{\text{H}}\right)\right\} + \frac{1}{2}\mathcal{E}\left\{\log\det\left(\mathbf{I}_N + P_2\mathbf{D}_1^{-1}\mathbf{H}_{21}\mathbf{H}_{21}^{\text{H}}\right)\right\}$$

(5.43)

where $\mathbf{H}_{12} = \mathbf{H}_2\mathbf{G}\mathbf{H}_1$ is the equivalent (diagonal) channel from "left to right" and $\mathbf{H}_{21} = \mathbf{H}_1^{\text{T}}\mathbf{G}\mathbf{H}_2^{\text{T}}$ the equivalent (diagonal) channel from "right to left". The orthogonalize-and-forward gain matrix $\mathbf{G}$ contains the elements of $\mathbf{g}$, which is the projection of $\mathbf{g}_\infty$ given in Section 5.1.3 on the nullspace of $\mathbf{H}_{\text{int,bi}}$. The diagonal matrices $\mathbf{D}_1$ and $\mathbf{D}_2$ contain the diagonal elements of the noise covariance matrices

$$\mathbf{R}_1 = \mathcal{E}\left\{\tilde{\mathbf{n}}_1\tilde{\mathbf{n}}_1^{\text{H}}\right\} = \sigma_3^2\mathbf{H}_1^{\text{T}}\mathbf{G}\mathbf{G}^{\text{H}}\mathbf{H}_1^* + \sigma_1^2\mathbf{I}_N,$$

(5.44)

where $\tilde{\mathbf{n}}_1 = \mathbf{H}_1^{\text{T}}\mathbf{G}\mathbf{n}_3 + \mathbf{n}_1$ is the overall noise at the terminals in $\mathcal{T}_1$ and

$$\mathbf{R}_2 = \mathcal{E}\left\{\tilde{\mathbf{n}}_2\tilde{\mathbf{n}}_2^{\text{H}}\right\} = \sigma_3^3\mathbf{H}_2\mathbf{G}\mathbf{G}^{\text{H}}\mathbf{H}_2^{\text{H}} + \sigma_2^2\mathbf{I}_N$$

(5.45)

respectively. Since the noise samples at the nodes in $\mathcal{T}_1$ and $\mathcal{T}_2$ are correlated we look also at the case of cooperative nodes in $\mathcal{T}_1$ and $\mathcal{T}_2$ with joint decoding of data streams (i.e., two users each equipped with multiple colocated antennas). The achievable rate is then

$$R_{\text{sum,joint}}^{\text{OF}} = \frac{1}{2}\mathcal{E}\left\{\log\det\left(\mathbf{I}_N + P_1\mathbf{R}_2^{-1}\mathbf{H}_{12}\mathbf{H}_{12}^{\text{H}}\right)\right\} + \frac{1}{2}\mathcal{E}\left\{\log\det\left(\mathbf{I}_N + P_2\mathbf{R}_1^{-1}\mathbf{H}_{21}\mathbf{H}_{21}^{\text{H}}\right)\right\}$$

(5.46)

where $\mathbf{g}$ is the projection of $\mathbf{g}_\infty$ on the nullspace of $\mathbf{H}_{\text{int}}$ (not on $\mathbf{H}_{\text{int,bi}}$), since the back-propagating self-interference $\mathbf{x}_2$ and $\mathbf{x}_1$ in (5.39) and (5.40) may be canceled at the corresponding multi-antenna terminals. Hence, for multiple antenna nodes the minimum relay configuration remains unaltered for the bidirectional orthogonalize-and-forward relaying scheme. Clearly, $R_{\text{sum,joint}}^{\text{OF}} \geq R_{\text{sum,sep}}^{\text{OF}}$, since the noise samples are correlated at the different users, but the rate loss due to separate decoding is small as the numerical examples in Section 5.4 show.

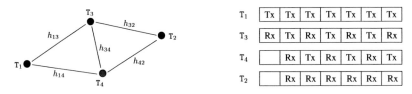

Fig. 5.3: Two-path relaying with alternating relays

5.3 Increasing the Spectral Efficiency with Two-path Relaying

In the protocol discussed in the previous section we required a bidirectional traffic between T_1 and T_2 in order to circumvent the loss in spectral efficiency. For the second protocol we assume a unidirectional traffic model but that two half-duplex relays assist in the communication and that no direct connection between terminals T_1 and T_2 is available. Transmission of messages from a source T_1 to a destination T_2 is done via two relays T_3 and T_4, which may not receive and transmit simultaneously. A message is transmitted in two time slots. In the first slot the source transmits the message to relay T_3 or T_4 and in the second time slot the message is forwarded to the destination, see Fig.5.3. The length of one time slot is equal to the length of one codeword (frame) and is NT, where T is the sampling interval and N the number of symbols in each frame. In odd time slots, $k = 1, 3, 5, \ldots$, relay T_3 receives and T_4 transmits. (Except for $k = 1$, where T_4 does not transmit), whereas in even time slots, $k = 2, 4, 6, \ldots$, it is the other way around. This cooperation protocol avoids the pre-log factor $\frac{1}{2}$ since the source transmits a new message in every time slot and has not to be *silent* in each second time slot. However, since the relays do not operate in orthogonal channels (as in [115]), there will be interference between T_3 and T_4 and it is not clear *a priori* whether this inter-relay interference cancels the gain achieved by the increased pre-log factor.

We will consider both *slow fading* and *fast fading*. In the case of slow fading we assume that all channel gains in the two-relay network remain constant during one time slot (codeword). In the case of fast fading we assume that the channels change i.i.d. from symbol to symbol.

5.3.1 Two-path Relaying for Fast Fading Channels

Assume that a sequence of K messages is to be transmitted. In time slot $k \in \{1, 2, \ldots, K\}$ the source T_1 chooses randomly a message (index) $M[k] \in \{1, 2, \ldots, 2^{NR[k]}\}$ according to a uniform distribution with $R[k]$ being the achievable rate for frame k. The message $M[k]$ is then mapped to a codeword $x_1[k] = (x_1[k, 1], x_1[k, 2], \ldots, x_1[k, N])$ of length N where the symbols $\{x_1[k, n]\}_{k,n}$ are i.i.d. according to $\mathcal{CN}(0, P_1)$ with P_1 being the average transmit power of the source. The channel gain between node i and node j at the discrete time $[k, n] := (kN + n)\, T$ is denoted as $h_{ij}[k, n]$, with $\mathcal{E}\left\{|h_{13}|^2\right\} = \mathcal{E}\left\{|h_{14}|^2\right\} = \nu_1^2$, $\mathcal{E}\left\{|h_{32}|^2\right\} = \mathcal{E}\left\{|h_{42}|^2\right\} = \nu_2^2$ and $\mathcal{E}\left\{|h_{34}|^2\right\} = \nu_{34}^2$. Due to notational simplicity, we assume channel reciprocity for the inter-relay channel and equal fading variances from the source to both relays and equal fading variances from both relays to the destination, respectively. We assume that $\{h_{ij}[k, n]\}_{k,n}$ are independent, stationary and ergodic fading processes. The source T_1 is aware of the *fading distribution* of the channel gains in the network but not of the *fading realizations*. The destination knows the *fading realizations* of all channel gains in the network. For the relay the assumptions about channel knowledge vary with the protocols (AF, DF).

5.3.1.1 Amplify-and-forward

The receive signal of relay T_p at time instant $(kN + n)\, T$ is given as

$$y_p[k, n] = h_{1p}[k, n]x_1[k, n] + n_p[k, n] + \sum_{i=1}^{k-1}\left(h_{1q}[k - i, n]x_1[k - i, n] + n_q[k - i, n]\right)f_i[k, n]$$

(5.47)

with $p = 4 - \mathrm{mod}\,(k, 2) \in \{3, 4\}$ and $q = 4 - \mathrm{mod}\,(k - i, 2) \in \{3, 4\}$[6] where

$$f_i[k, n] := \prod_{j=1}^{i} h_{34}[k - j, n]g[k - j]$$

(5.48)

denotes the *inter-relay interference factor*. The relay noise samples $\{n_i[k, n]\}_{k,n}$, $i \in \{3, 4\}$ are i.i.d. according to $\mathcal{CN}(0, \sigma_\mathrm{R}^2)$ and the destination noise samples $\{n_2[k, n]\}_{k,n}$ are i.i.d. according to $\mathcal{CN}(0, \sigma_2^2)$. The transmit signal of relay T_p is a scaled version of its received signal: $t_p[k + 1, n] = g[k]y_p[k, n]$, where $g[k]$ is the relay scaling coefficient and for $k = 2, 3, 4, \ldots$ chosen as

$$g^2[k] = \frac{P_\mathrm{R}}{\frac{1}{N}\sum_{n=1}^{N}|y_p[k, n]|^2} \approx \frac{P_\mathrm{R}}{P_1\nu_1^2 + P_\mathrm{R}\nu_{34}^2 + \sigma_\mathrm{R}^2} := g^2$$

(5.49)

and in the first time slot $k = 1$

$$g^2[1] = \frac{P_\mathrm{R}}{\frac{1}{N}\sum_{n=1}^{N}|y_3[1, n]|^2} \approx \frac{P_\mathrm{R}}{P_1\nu_1^2 + \sigma_\mathrm{R}^2} \geq g^2$$

(5.50)

---

[6]p and q is used to differentiate between relay T_3 and T_4.

where $P_3 = P_4 = P_R$ is the average transmit power of each relay. The approximations in (5.49) and (5.50) are exact for $N \to \infty$ by the law of large numbers. Destination T_2 observes at $((k + 1)N + n)T$ the signal

$$y_2[k + 1, n] = h_{p2}[k + 1, n]gy_p[k, n] + n_2[k + 1, n] \qquad (5.51)$$

where $y_p[k, n]$ is given in (5.47). To decode $x_1[k]$ from (5.51) the destination receiver first subtracts the previously decoded codewords $x_1[k-1], \ldots, x_1[1]$ from the received frame $y_2[k+1]$, because these codewords appear as accumulated inter-relay interference at the destination. However, the influence of the codewords transmitted several time slots before k is weak since they were attenuated several times by the inter-relay channel h_{34}, which acts as forgetting factor for the decoding process, see (5.47) and (5.48). After perfect cancelation of m previously decoded codewords $x_1[k - 1], x_1[k - 2], \ldots, x_1[k - m]$ the destination signal is given by

$$y_2^{(m)}[k + 1, n] = h_{p2}[k + 1, n]gh_{1p}[k, n]\left(x_1[k, n] + \sum_{i=m+1}^{k-1} h_{1q}[k - i, n]x_1[k - i, n]f_i[k, n] \right.$$
$$\left. + \sum_{i=0}^{k-1} n_q[n - i, n]f_i[k, n] \right) + n_2[k + 1, n] \qquad (5.52)$$

where $f_i[k, n] := 1$ for $i = 0 \; \forall k, n$. For $m = k - 1$ all previously transmitted codewords are canceled (*full interference cancelation*). For $m = 0$ all codewords up to $x_1[k - 1]$ appear as inter-frame interference when $x_1[k]$ is decoded. For $0 < m < k - 1$ only the last m transmitted codewords are canceled and $x_1[1], x_1[2], \ldots, x_1[k - m - 1]$ remain as interference terms (*partial interference cancelation*). The ergodic rate in time slot $k + 1$ measured in b/s/Hz follows as

$$R[k + 1] = \mathcal{E}\left\{ \log\left(1 + \frac{P_1|h_{p2}gh_{1p}|^2}{\sigma_2^2 + |h_{p2}g|^2\left(P_1I_1[k] + \sigma_R^2I_2[k]\right)} \right) \right\} \qquad (5.53)$$

with $I_1[k]$ denoting the accumulated inter-frame interference given by

$$I_1[k] = \sum_{i=m+1}^{k-1} |h_{1p}|^2|f_i[k]|^2 \qquad (5.54)$$

and $I_2[k]$ the accumulated noise interference given by

$$I_2[k] = \sum_{i=0}^{k-1} |f_i[k]|^2. \tag{5.55}$$

Note that $f_i[k]$ models the inter-relay interference factor as random variable whose statistics depends on k whereas $f_i[k,n]$ denotes its realization in time slot k at symbol time n[7]. After the first time slot, i.e., for $k = 1$, we have $I_1[1] = 0$ and $I_2[1] = 1$. Clearly, $R[1] = 0$ because after transmission of the first frame no signal is received by the destination yet. The expectation is taken with respect to the statistics of h_{1p}, h_{p2} for $p \in \{3,4\}$ and h_{34} and depends on the channel model that is used for the fading variables. After transmission of a sequence of K messages we get the average rate

$$\overline{R}_K = \frac{1}{K+1} \sum_{k=1}^{K} R[k+1] \tag{5.56}$$

$$\geq \frac{K}{K+1} R[K+1] \tag{5.57}$$

$$\geq \frac{K}{K+1} \lim_{k \to \infty} R[k] \tag{5.58}$$

where the pre-log is $\frac{K}{K+1} \approx 1$ for large K. The inequalities (5.57) and (5.58) are motivated by the observation that the average interference power between the relays is upper bounded by the average relay transmit power, i.e, the rate does not diminish as k grows large. In order to see this we look at a lower bound on (5.53):

$$R[k+1] > \mathcal{E} \left\{ \log \left(\frac{P_1 |h_{p2} g h_{1p}|^2}{\sigma_2^2 + |h_{p2} g|^2 \left(P_1 I_1[k] + \sigma_R^2 I_2[k] \right)} \right) \right\} \tag{5.59}$$

$$> \mathcal{E} \left\{ \log \left(P_1 |h_{p2} g h_{1p}|^2 \right) \right\} - \sigma_2^2 - \mathcal{E} \left\{ |h_{p2} g|^2 \left(P_1 I_1[k] + \sigma_R^2 I_2[k] \right) \right\} \tag{5.60}$$

$$\overset{!}{>} 0. \tag{5.61}$$

From that it follows

$$\mathcal{E} \left\{ \log \left(P_1 |h_{p2} g h_{1p}|^2 \right) \right\} > \sigma_2^2 + g^2 \mathcal{E} \left\{ |h_{p2}|^2 \right\} \mathcal{E} \left\{ \left(P_1 I_1[k] + \sigma_R^2 I_2[k] \right) \right\} \tag{5.62}$$

$$\frac{2^{\mathcal{E} \left\{ \log \left(P_1 |h_{p2} g h_{1p}|^2 \right) \right\}} - \sigma_2^2}{g^2 \nu_2^2} > P_1 \overline{I}_1[k] + \sigma_R^2 \overline{I}_2[k] = P_R \nu_{34}^2 \tag{5.63}$$

---

[7]Similar for the channels: h_{ij} is the random variable whose statistics remain the same for all time and $h_{ij}[k,n]$ its realization at $(kN + n)T$.

where

$$\bar{I}_1[k] = \sum_{i=m+1}^{k-1} \mathcal{E}\left\{|f_i[k]|^2\right\} \tag{5.64}$$

$$\bar{I}_2[k] = \sum_{i=0}^{k-1} \mathcal{E}\left\{|f_i[k]|^2\right\}. \tag{5.65}$$

We evaluate the expectation in (5.63) for Gaussian fading first hop and second hop channels with zero mean and unit variance (double Rayleigh fading channel), i.e., $\nu_1^2 = \nu_2^2 = 1$:

$$\frac{c \cdot P_1 \cdot g^2 - \sigma_2^2}{g^2} \approx cP_1 > P_{\mathrm{R}}\nu_{34}^2 \tag{5.66}$$

where $c = 1.12$. We obtain the following condition for the relay transmit power

$$P_{\mathrm{R}} < \frac{1.12 \cdot P_1}{\nu_{34}^2}. \tag{5.67}$$

This means that as long as the relay transmit power fulfils (5.67) the lower bound is larger than zero and therefore also the rate (5.53). The disadvantage of signaling according to (5.53) is that the source has to adapt the rate for each frame. However, the lower bounds in (5.57) and (5.58) suggest to use a fixed-rate scheme at the source, either $R[K+1]$ or $\lim_{k \to \infty} R[k]$. By using $\lim_{k \to \infty} R[k]$ the rate is independent of the number of messages K to be transmitted. In order to simplify the computation of $\lim_{k \to \infty} R[k]$ we lower bound the rate (5.53) in a different way from the previous lower bound. For $k = 2, 3, \ldots, K$ it is

For $k = 2, 3, \ldots, K$ it is

$$\begin{aligned}
R[k+1] &= \mathcal{E}\left\{\log\left(\frac{\sigma_2^2 + |h_{p2}g|^2\big(P_1 I_1[k] + \sigma_{\mathrm{R}}^2 I_2[k]\big) + P_1|h_{p2}gh_{1p}|^2}{\sigma_2^2 + |h_{p2}g|^2\big(P_1 I_1[k] + \sigma_{\mathrm{R}}^2 I_2[k]\big)}\right)\right\} \tag{5.68} \\
&\geq \mathcal{E}\left\{\log\left(\frac{\sigma_2^2 + |h_{p2}g|^2\big(P_1 I_{1,k'}[k] + \sigma_{\mathrm{R}}^2 I_{2,k'}[k]\big) + P_1|h_{p2}gh_{1p}|^2}{\sigma_2^2 + |h_{p2}g|^2\big(P_1 I_1[k] + \sigma_{\mathrm{R}}^2 I_2[k]\big)}\right)\right\} \tag{5.69} \\
&\geq \mathcal{E}\left\{\log\left(P_1|h_{p2}gh_{1p}|^2 + \sigma_2^2 + |h_{p2}g|^2\big(P_1 I_{1,k'}[k] + \sigma_{\mathrm{R}}^2 I_{2,k'}[k]\big)\right)\right\} \\
&\quad - \log\left(\sigma_2^2 + \nu_2^2 g^2\big(P_1 \bar{I}_1[k] + \sigma_{\mathrm{R}}^2 \bar{I}_2[k]\big)\right) \tag{5.70} \\
&= R_{\mathrm{low}}[k+1]
\end{aligned}$$

127

with

$$I_{1,k'}[k] = \sum_{i=m+1}^{k'} |h_{1q}|^2 |f_i[k]|^2 \tag{5.71}$$

$$I_{2,k'}[k] = \sum_{i=0}^{k'} |f_i[k]|^2 \tag{5.72}$$

and

$$\overline{I}_1[k] = \sum_{i=m+1}^{k-1} \mathcal{E}\left\{|f_i[k]|^2\right\} = \frac{u^{m+1} - u^k}{1 - u} \tag{5.73}$$

$$\overline{I}_2[k] = \sum_{i=0}^{k-1} \mathcal{E}\left\{|f_i[k]|^2\right\} = \frac{1 - u^k}{1 - u} \tag{5.74}$$

where $u = g^2 \nu_{34}^2$. The first inequality (5.69) follows due to

$$I_1[k] \geq I_{1,k'}[k] = \sum_{i=m+1}^{k'} |h_{1q}|^2 |f_i[k]|^2 \tag{5.75}$$

and

$$I_2[k] \geq I_{2,k'}[k] = \sum_{i=0}^{k'} |f_i[k]|^2 \tag{5.76}$$

for $k' < k - 1$. The second inequality (5.70) follows by applying Jensen's inequality [74] on the second log – term. For $k \to \infty$ and $q < 1$ we get

$$\lim_{k \to \infty} \overline{I}_1[k] = \frac{u^{m+1}}{1 - u} \tag{5.77}$$

and

$$\lim_{k \to \infty} \overline{I}_2[k] = \frac{1}{1 - u} \tag{5.78}$$

and the lower bound (5.70) becomes independent of the actual frame number k. Note that for a stationary inter-relay channel h_{34} the statistics of $I_{1,k'}[k]$ and $I_{2,k'}[k]$ become independent of k (but dependent on k'). Numerical results in Section 5.4 show that fixed-rate signaling according to $\lim_{k \to \infty} R_{\text{low}}[k]$ or $R_{\text{low}}[K+1]$ in each frame induces only a small loss compared to variable-rate signaling according to (5.53), but has the advantage that the source does not have to adapt the rate for each frame. The parameter k' can be used to improve the lower bound:

the larger k' the better the lower bound, but the more involved becomes the evaluation of the expectation of the first log-term in (5.70). Note that the purpose of the lower bound in (5.70) is not to give qualitative insights into the performance of the two-path AF protocol but to show that one could use a fixed-rate signaling scheme based on such a lower bound. The lower bound works fine when the relay power is not too large. For the numerical examples in Section 5.4 we have chosen the relay power to be equal to the source power. Another possibility to obtain a rate for fixed-rate signaling would be to determine the ergodic rate (5.53) for a specific channel model and then to compute (5.57) or (5.58). However, for most channel models it will not be possible to do this analytically.

5.3.1.2 Decode-and-forward

At the discrete time $[k, n] := (kN + n) T$ the receive signal of relay T_p with $p = 4 - \text{mod} (k, 2) \in \{3, 4\}$ is given as

$$y_p[k, n] = h_{1p}[k, n]x_1[k, n] + h_{34}[k, n]x_1[k - 1, n] + n_p[k, n]. \tag{5.79}$$

Note that relay T_p is interested only in $x_1[k, n]$ since $x_1[k - 1, n]$ is forwarded by the other relay to the destination, and hence $x_1[k - 1, n]$ appears as interference at relay T_p. In order to simplify the exposition we assume that both channels in the first hop, i.e., h_{13} and h_{14} have the same channel statistics. Similarly, the channels in the second hop, i.e., h_{32} and h_{42} have the same channel statistics. From this it follows that the source uses the same code rate in each time slot, i.e., it does not matter to which relay it currently has to transmit. Instead of treating (5.79) as interference channel we rather look at it as a multiple access channel, i.e., source T_1 and relay T_4 (T_3) transmit simultaneously to relay T_3 (T_4) which has to decode the signal from the source T_1 but may also decode the signal from relay T_4 (T_3) if it helps to increase the overall transmission rate. In the following we propose three different decoding strategies at the relay T_p.

Strategy a). The relay decodes $x_1[k, n]$ treating $h_{34}[k, n]x_1[k - 1, n]$ as interference. The achievable rate is given by

$$R_{\mathrm{a}} = \min \left(C \left(\frac{P_1 |h_{1p}|^2}{\sigma_{\mathrm{R}}^2 + P_{\mathrm{R}} |h_{34}|^2} \right), C \left(\frac{P_{\mathrm{R}} |h_{p2}|^2}{\sigma_2^2} \right) \right) \tag{5.80}$$

where $C(x) = \mathcal{E} \{\log(1 + x)\}$. The first term in (5.80) determines the maximum rate in the first hop (source T_1 to relay T_p) when the inter-relay signal is treated as interference and the second term denotes the maximum rate in the second hop (relay T_p to destination T_2).

Strategy b). The relay decodes first $x_1[k-1, n]$ treating $h_{34}[k, n]x_1[k, n]$ as interference, subtracts $h_{34}[k, n]x_1[k-1, n]$ from the received signal $y_p[k, n]$, then decodes $x_1[k, n]$ interference-free. The achievable rate with this strategy is

$$R_b = \min\left(C\left(\frac{P_1|h_{1p}|^2}{\sigma_R^2}\right), C\left(\frac{P_R|h_{p2}|^2}{\sigma_2^2}\right), C\left(\frac{P_R|h_{34}|^2}{\sigma_R^2 + P_1|h_{1p}|^2}\right)\right). \tag{5.81}$$

The first term in (5.81) denotes the maximum rate in the first hop, when there is no interference from the other relay, the second term is again the maximum rate in the second hop. The third term is the maximum rate from one relay to the other relay when the source signal is treated as interference.

Strategy c). The relay decodes $x_1[k-1, n]$ and $x_1[k, n]$ jointly[8] and then forwards $x_1[k, n]$ only. The achievable rate is

$$R_c = \min\left(C\left(\frac{P_1|h_{1p}|^2}{\sigma_R^2}\right), C\left(\frac{P_R|h_{34}|^2}{\sigma_R^2}\right), \frac{1}{2}C\left(\frac{P_1|h_{1p}|^2 + P_R|h_{34}|^2}{\sigma_R^2}\right), C\left(\frac{P_R|h_{p2}|^2}{\sigma_2^2}\right)\right). \tag{5.82}$$

Due to the assumption that h_{13} and h_{14} obey the same channel statistics, we know that the rate used by the source in even and odd time slots is the same, hence the first three terms in (5.82) denote the multiple access rate region with symmetric "user" rates. The third term multiplied by the factor two is the sum rate that can be decoded reliably at relay T_p. Simultaneously, the rate has to be smaller than the "single user" capacities from source to relay and from relay to relay (first and second term in (5.82)) as well as smaller than the capacity in the second hop (fourth term in (5.82)).

Strategy a) works well when the inter-relay channel is not too strong. Strategy b) works well, when the inter-relay channel is strong, since the rate in (5.81) is not limited by the third expression and the relay may decode the new message coming from the source interference-free. Strategy c) works good, for moderately to strong inter-relay channels h_{34}. The reason is, that in this case the interference cancelers in strategies a) and b) have to cope with strong interference which leads to small achievable rates. Fig. 5.4 shows the capacity region when the channels h_{1p} and h_{34} have the same variance. We see that strategies a) and b) are operating far from the optimal boundary of the region and the strategy c) is able to considerably improve the rate in such a situation.

---

[8]Which can be done by a maximum-likelihood or typical sequences decoder.

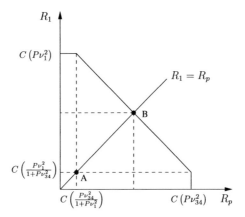

Fig. 5.4: Capacity region of the multiple access channel with T_1 and T_4 (T_3) being "users" and $T3$ (T_4) being the "base station". The region is depicted for $\mathcal{E}\{|h_{13}|^2\} = \mathcal{E}\{|h_{14}|^2\} = \nu_1^2$ and $\mathcal{E}\{|h_{34}|^2\} = \nu_{34}^2$. Point A is achievable by strategies a) and b) and point B is achievable by strategy c).

5.3.2 Two-path Relaying for Slow Fading Channels

In the previous section we looked at two-path relaying with AF and DF relays for fast-fading channels. The DF case can easily be applied to slow fading channels. One has simply to replace the function $C(x) = \mathcal{E}\{\log(1+x)\}$ by $C(x) = \log(1+x)$ in the equations (5.80)–(5.82) for strategies a), b) and c). For all strategies, outage occurs when the rate R chosen by the source is smaller than the expressions given in (5.80)–(5.82).

The AF two-path protocol is trickier when we consider slow fading. Recall that we required the destination to cancel the accumulated inter-relay interference up to time $k-1$ before decoding the new codeword in time k. However, in a slow fading environment, it may be that the mutual information of the channel is lower than the rate chosen by the source and therefore, it is not always possible to decode, i.e., outage can occur. But this also means, that when the destination wants to decode the codeword $x[k]$, it possibly cannot subtract all interference terms $x[1], x[2], \ldots, x[k-1]$ since some of the codewords have possibly not been decoded properly due to outage. In this section we propose another encoding and decoding strategy which suits better to slow fading channels.

Assume that in time slot k the source T_1 chooses randomly a message (index) $M[k] \in \{1, 2, \ldots, 2^{NR}\}$ according to a uniform distribution with R being the information rate. The

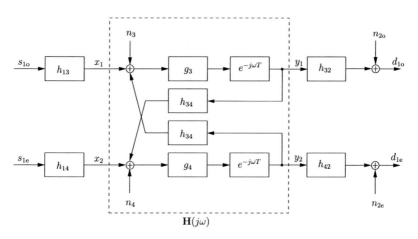

Fig. 5.5: Linear and time-invariant system model of two-path relaying with slow fading

message $M[k]$ is then mapped to a codeword[9] $s_1[k] = (s_1[k, 1], s_1[k, 2], \ldots, s_1[k, N])$ of length N where the symbols $\{s_1[k, n]\}_{k,n}$ are i.i.d. according to $\mathcal{CN}(0, P_1)$ with P_1 being the average transmit power of the source. The channel gain between node i and node j at the discrete time $[k, n] := (kN + n)T$ is denoted as $h_{ij}[k]$ for all n, i.e., the channel gain remains constant during one codeword and changes independently over time slots k. Again, due to notational simplicity, we assume channel reciprocity for the inter-relay channel and equal fading variances from the source to both relays and equal fading variances from both relays to the destination, respectively, i.e., $\mathcal{E}\{|h_{13}|^2\} = \mathcal{E}\{|h_{14}|^2\} = \nu_1^2$, $\mathcal{E}\{|h_{32}|^2\} = \mathcal{E}\{|h_{42}|^2\} = \nu_2^2$ and $\mathcal{E}\{|h_{34}|^2\} = \nu_{34}^2$. The source T_1 is aware of the *fading distribution* of the channel gains in the network but not of the *fading realizations*. The destination knows the *fading realizations* of all channel gains in the network. Relay T_3 knows the fading realizations of h_{13} and of h_{34} and relay T_4 knows the fading realizations of h_{14} and of h_{34}. Since the channels change slowly, we assume here, that the relays know the channels from which they receive. Therefore, the relay gains g_3 and g_4, see Fig.5.5 can be chosen based on this channel knowledge.

We model the two-path relaying system depicted in Fig.5.3 as a linear time-invariant system with feedback, see Fig.5.5. The signals s_{1e} and s_{1o} model the even and odd states of the source,

---

[9]In this section we change the notation slightly: the source symbols are now $s[k]$ instead of $x[k]$ and the destination receive symbols are $d[k]$ instead of $y_2[k]$.

i.e.,

$$s_{1e}[k] = \begin{cases} s_1[k], & k=2,4,6,\dots; \\ 0, & k=1,3,5,\dots \end{cases} \tag{5.83}$$

and

$$s_{1o}[k] = \begin{cases} 0, & k=2,4,6,\dots; \\ s_1[k], & k=1,3,5,\dots \end{cases} \tag{5.84}$$

The frequency-domain input-output relation between the signals $\mathbf{x} = (x_1, x_2)^T$ and the signals $\mathbf{y} = (y_1, y_2)^T$ (see Fig.5.5) is given by

$$\mathbf{y} = \mathbf{H}(j\omega)\mathbf{x} + \mathbf{H}(j\omega)\mathbf{n}_{34} \tag{5.85}$$

where $\mathbf{n}_{34} = (n_3, n_4)^T$ and

$$\mathbf{H}(j\omega) = \begin{bmatrix} H_{11}(j\omega) & H_{12}(j\omega) \\ H_{21}(j\omega) & H_{22}(j\omega) \end{bmatrix}. \tag{5.86}$$

In the following we determine the transfer function $\mathbf{H}(j\omega)$. First we set $x_1 = 0$, which leads to

$$y_1 = H_{11}(j\omega)x_1 \tag{5.87}$$
$$y_2 = H_{21}(j\omega)x_1. \tag{5.88}$$

In a general feedback system, the transfer function is determined from the equation

$$y = H_f x + H_b y \tag{5.89}$$

where H_f is the forward part of the signal transmission and H_b the feedback part of the signal transmission. It follows that

$$\frac{y}{x} = \frac{H_f}{1 - H_b} = H_{ij}(j\omega). \tag{5.90}$$

From (5.90) and Fig.5.5 we obtain

$$H_{11}(j\omega) = \frac{g_3 e^{-j\omega T}}{1 - h_{34}^2 g_3 g_4 e^{-j\omega 2T}} \tag{5.91}$$

$$H_{21}(j\omega) = \frac{h_{34} g_3 g_4 e^{-j\omega 2T}}{1 - h_{34}^2 g_3 g_4 e^{-j\omega 2T}}. \tag{5.92}$$

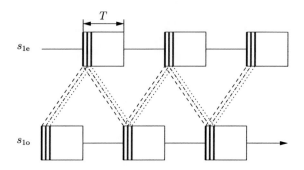

Fig. 5.6: Interference mechanism in two-path relaying: samples of same color do interfere with each other. Samples with different colors do not interfere.

Similarly, for $x_2 = 0$, we obtain

$$H_{12}(j\omega) = \frac{h_{34}g_3g_4e^{-j\omega 2T}}{1 - h_{34}^2g_3g_4e^{-j\omega 2T}} \tag{5.93}$$

$$H_{22}(j\omega) = \frac{g_4e^{-j\omega T}}{1 - h_{34}^2g_3g_4e^{-j\omega 2T}}. \tag{5.94}$$

The transfer matrix $\mathbf{H}(j\omega)$ is therefore

$$\mathbf{H}(j\omega) = \frac{1}{1 - A} \left[\begin{array}{cc} g_3e^{-j\omega T} & h_{34}g_3g_4e^{-j\omega 2T} \\ h_{34}g_3g_4e^{-j\omega 2T} & g_4e^{-j\omega T} \end{array} \right] \tag{5.95}$$

where $A = h_{34}^2g_3g_4e^{-j\omega 2T}$. Note that the interference in $\mathbf{y}$ is only within correspondent samples of adjacent blocks (see Fig.5.6), i.e., we can assume the block length $N = 1$ without loss of generality. We introduce the diagonal matrices

$$\mathbf{D}_i = \left[\begin{array}{cc} 1 & 0 \\ 0 & e^{-j\omega T} \end{array} \right] \tag{5.96}$$

and

$$\mathbf{D}_o = \left[\begin{array}{cc} e^{-j\omega T} & 0 \\ 0 & 1 \end{array} \right]. \tag{5.97}$$

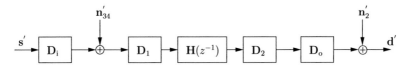

Fig. 5.7: Equivalent system model of two-path relaying with time aligned signals.

Further we define the diagonal matrices

$$\mathbf{D}_1 = \begin{bmatrix} h_{13} & 0 \\ 0 & h_{14} \end{bmatrix} \tag{5.98}$$

and

$$\mathbf{D}_2 = \begin{bmatrix} h_{32} & 0 \\ 0 & h_{42} \end{bmatrix}. \tag{5.99}$$

Now we are ready to provide an equivalent system model with inputs and outputs temporarily aligned:

$$\mathbf{d}' = \mathbf{s}' \mathbf{D}_i \mathbf{D}_1 \mathbf{H} \mathbf{D}_2 \mathbf{D}_o + \mathbf{n}'_{34} \mathbf{D}_1 \mathbf{H} \mathbf{D}_2 \mathbf{D}_o + \mathbf{n}'_2 \tag{5.100}$$

where $\mathbf{s}'$, $\mathbf{d}'$, $\mathbf{n}'_{34}$ and $\mathbf{n}'_2$ are temporarily aligned versions of $\mathbf{s}$, $\mathbf{d}$, $\mathbf{n}_{34}$ and $\mathbf{n}_2$, respectively and $\mathbf{n}_2 = (n_{2e}, n_{2o})^{\mathrm{T}}$. The equivalent system model is illustrated in Fig.5.7. The mutual information between $\mathbf{s}'$ and $\mathbf{d}'$ for given channel realizations h_{13}, h_{14}, h_{32} and h_{42} is given by

$$I(\mathbf{s}'; \mathbf{d}') = \frac{1}{4\pi} \int\limits_{-\pi}^{\pi} \log\left(\det\left(\mathbf{I}_2 + P\mathbf{R}(j\omega)^{-1} \mathbf{H}_{\mathrm{eq}}(j\omega) \mathbf{H}_{\mathrm{eq}}(j\omega)\right)\right) \mathrm{d}\omega \tag{5.101}$$

where

$$\mathbf{H}(j\omega)_{\mathrm{eq}} = \mathbf{D}_i \mathbf{D}_1 \mathbf{H}(j(\omega)) \mathbf{D}_2 \mathbf{D}_o \tag{5.102}$$

is the equivalent channel transfer function and

$$\mathbf{R}(j\omega) = \mathbf{I}_2 + (\mathbf{D}_1 \mathbf{H}(j(\omega)) \mathbf{D}_2 \mathbf{D}_o) (\mathbf{D}_1 \mathbf{H}(j(\omega)) \mathbf{D}_2 \mathbf{D}_o)^{\mathrm{H}} \tag{5.103}$$

is the noise covariance matrix. Note that we assume all noise variances to be one. In order to compute the CDF of (5.101) or the average rate we will solve the integral in (5.101) numerically and show the results in the next section.

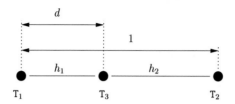

Fig. 5.8: Simplified relay network for two-way relaying

5.4 Simulation Results

5.4.1 Two-way Relaying

We evaluate the achievable rates of the relaying schemes described in Section 5.2 by Monte Carlo simulations. We consider a linear one-dimensional network geometry as in Fig.5.8, where the distance between terminal T_1 and T_2 is normalized to one. The channel gains are modeled as $h_i = \frac{\xi_i}{d_i^{\alpha/2}}$ with i.i.d. $\xi_i \sim \mathcal{CN}(0,1)$ (Rayleigh fading), where d is the normalized distance between terminal T_1 and relay T_3, and $\alpha = 3$ the path loss exponent. The noise variances at the terminals are chosen as $\sigma_1^2 = \sigma_2^2 = \sigma_3^2 = 1$ and the transmit powers $P_1 = P_2 = P_3 = 10$.

In Fig.5.9 we compare the sum-rate of the two-way AF and two-way DF protocol with their one-way counterparts and the cut-set upper bound [74] applied to the two-way half-duplex relay channel with no direct link. We observe that both protocols do not achieve the cut-set upper bound on the sum-rate which is given by

$$
C_{\text{sum}}^{\text{u}} = \max_{0 \leq \beta \leq 1} \left(\min \left(C\left(P_1|h_1|^2\right), C\left(\beta P_3|h_2|^2\right) \right) \right.
$$
$$
\left. + \min \left(C\left(P_2|h_2|^2\right), C\left((1-\beta)\beta P_3|h_1|^2\right) \right) \right). \tag{5.104}
$$

Both two-way protocols, AF and DF, achieve a sum-rate that is substantially larger than the rates of their one-way counterparts. In the one-way AF and DF scenario both protocols achieve the highest rate when the relay is exactly in the middle and DF achieves a higher maximum than AF. For the two-way case the DF protocol is worse than the AF protocol when the relay is in the middle. The reason is that the DF scheme has to cope with a multiple-access channel and the maximum sum-rate is achieved here for an asymmetric channel situation, i.e., one terminal experiences a stronger channel gain than the other terminal.

In Fig.5.10 we compare the sum-rates of one-way and two-way OF relaying. We see that the

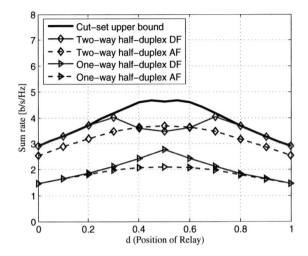

Fig. 5.9: Sum-rate for two-way half-duplex Af and DF relaying protocols

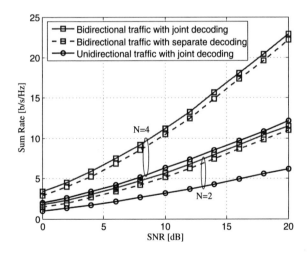

Fig. 5.10: Sum-rate for two-way half-duplex OF relaying protocol

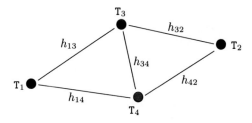

Fig. 5.11: Two-path Relaying

loss due to separate decoding is small, since the transmissions are orthogonalized and neglecting the noise correlations at the receiving nodes does not degrade the performance significantly. We observe that two-way OF relaying with $N = 2$ terminal pairs and $K = 5$ relays (minimum relay configuration) achieves almost the same sum-rate as one-way OF relaying with $N = 4$ terminal pairs and $K = 13$ relays.

5.4.2 Two-path Relaying

We evaluate the achievable rates of the relaying schemes described in Section 5.3 and depicted in Fig.5.11 again by Monte Carlo simulations. We assume i.i.d (in space and time) channel gains $h_{13}, h_{14} \sim \mathcal{CN}(0, \nu_1^2)$, $h_{32}, h_{42} \sim \mathcal{CN}(0, \nu_2^2)$ and $h_{34} \sim \mathcal{CN}(0, \nu_{34}^2)$. For simplicity we assume $\nu_1^2 = \nu_2^2 = 1$ (symmetric network). The noise variances are chosen to be $\sigma_R^2 = \sigma_2^2 = \sigma^2$ and the transmit powers $P_1 = P_3 = P_4 = P$. We simulated 5000 random channels for each value in Figs.5.12–5.15. The SNR is defined as $\text{SNR} = \frac{P}{\sigma^2}$.

In Fig.5.12 we see that for $\nu_{34}^2 = 0.5$ (inter-relay channel gain is 3dB weaker than the other channels) the two-path relaying protocol with AF relays and full cancelation of the accumulated inter-frame interference achieves an average rate $\overline{R}_K$ that is near to the rate of one full-duplex relay and outperforms clearly the case where only one half-duplex relay is used. For the lower bound (5.70) we have chosen $k' = 0$ and $I_1[k] = 0$ (full interference cancelation). Further we observe from Fig.5.12 that the performance loss of the fixed-rate schemes based on $\lim_{k\to\infty} R_{\text{low}}[k]$ or $R_{\text{low}}[K + 1]$ is small compared to the performance of the variable-rate scheme (5.53).

Fig.5.13 shows the achievable rate (5.53) for different variances ν_{34}^2 of the inter-relay channel gain and $P/\sigma^2 = 100$. When the inter-relay channel gain is not too strong, the two-path AF relaying strategy performs very well. For inter-relay channel gains that are considerably stronger than the source-relay and relay-destination channel gains the two-path strategy does not work well due to the accumulated noise interference of the two relays at the destination.

In Fig.5.14 we compare the impact of full and partial interference cancelation on the average rate for $P/\sigma^2 = 100$ and $\nu_{34}^2 = 1/2$. For the simulation the number of transmitted frames K was chosen to be 30. We see that after cancelation of about five to six previously transmitted codewords the performance is the same as with full cancelation of the inter-frame interference. The inter-relay channel acts as forgetting factor (5.48) and lessens the inter-frame interference caused by the previously transmitted codewords.

In Fig.5.15 we compare the rates achievable by two-path DF relaying for the three decoding strategies a), b) and c). The SNR is fixed at $P/\sigma^2 = 100$. We see that for strong inter-relay channels the relay should perform interference cancelation before decoding the message coming from the source (strategy b)) and for weak inter-relay channels it is better to treat the signal from the other relay as interference (strategy b)). For the case where the inter-relay channel is approximately equally strong as the first hop channels it's better to decode both signals jointly (strategy c)). We also plotted the rate of the two-path AF scheme for comparison. With an adaptive protocol which chooses based on the second-order statistics of the channels the best strategy it is possible to achieve a rate that is always above the rate of the single DF relay channel. Note that in all strategies the total network energy consumption per time slot is $2P$.

For the two-path relaying scheme with slow fading we evaluate the integral in (5.101) numerically. Fig.5.16 shows the CDF curves of the achievable rate (5.101) for 5000 random channel realizations h_{13}, h_{14}, h_{32} and h_{42}, each with variance 1. SNR is equal to 20 dB and the energy of the inter-relay channel varies from -10 dB to 10 dB. The relay gains have been chosen as

$$g_3 = \frac{P}{P|h_{13}|^2 + P|h_{34}|^2 + 1} \tag{5.105}$$

$$g_4 = \frac{P}{P|h_{14}|^2 + P|h_{34}|^2 + 1} \tag{5.106}$$

where P is the transmit power of each node. We compare the performance of the two-path protocol again with one single full-duplex relay and one single half-duplex relay (but with double transmit power $2P$). We see that the two-path protocol has about the same diversity order as the single relay channels, since each symbols goes either trough relay $T3$ or relay T_4 but not through both of them.

Fig.5.17 shows the average rate of the two-path protocol with same parameters as before. We see that for a inter-relay channel that is 10 dB weaker than the first-hop and second-hop channels the two-path protocol recovers the half-duplex pre-log loss completely. Even inter-relay channels

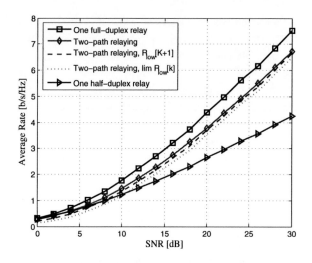

Fig. 5.12: Average rate vs. SNR for two-path half-duplex AF relaying protocol

with 10 dB higher channel energy than the first-hop and second-hop channels the two-path protocol performs better than the single half-duplex relay channel with power $2P$.

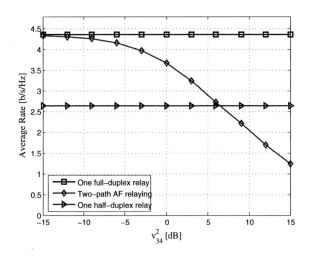

Fig. 5.13: Average rate vs. inter-relay channel gain for two-path half-duplex AF relaying protocol

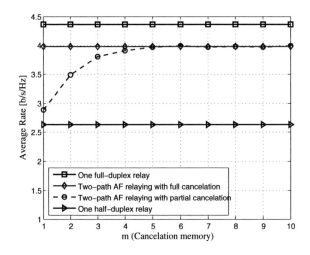

Fig. 5.14: Average rate vs. cancelation memory for two-path half-duplex AF relaying protocol

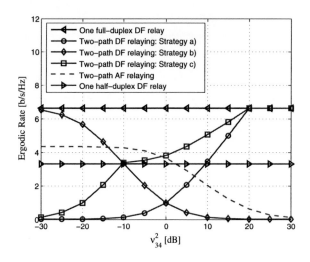

Fig. 5.15: Average rate vs. inter-relay channel gain for two-path half-duplex DF relaying protocol

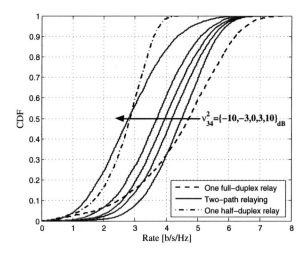

Fig. 5.16: CDF of achievable rates for the two-path protocol with slow fading. SNR is equal to 20 dB and the energy of the inter-relay channel varies from -10 dB to 10 dB.

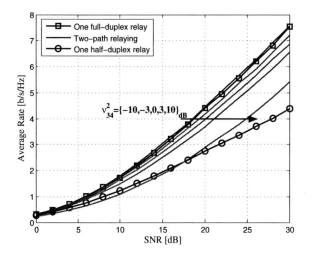

Fig. 5.17: Average rates for the two-path protocol with slow fading

6 Spectral Efficient Signaling in Rank-deficient MIMO Channels by the Use of Relays

6.1 Introduction

Multiple antennas at a transmitter and a receiver introduce spatial degrees of freedom into a wireless communication system. Space-time signal processing utilizes these degrees of freedom to boost link capacity and/or to enhance link reliability of multiple-input multiple-output (MIMO) communication systems. With *spatial multiplexing* one can increase the data rate without additional cost of bandwidth or power by transmitting data streams simultaneously over spatial sub-channels which are available in a rich scattering environment [10]. With space-time coding it is possible to mitigate the fading effects by utilizing the spatial diversity of the MIMO channel [12]. It is expected that future wireless broadband communication systems will operate beyond 5 GHz, for example Wireless Local Area Networks (WLANs) at 17 GHz (Hiperlan) or at 24/60 GHz (ISM bands). In higher frequency bands it is possible to accommodate a larger number of antennas in a given volume ("rich array") because the array size scales down with increasing frequency. Further on, the array gain of the system can compensate the path loss which is inversely proportional to the square of the frequency [92].

For zero-mean i.i.d. Gaussian channel coefficients the ergodic capacity of a MIMO channel with M transmit and N receive antennas scales linearly with $\min\{M, N\}$ compared to a corresponding single-input single-output (SISO) channel [9]. However, there is a major obstacle in the practical exploitation of MIMO technology: the capacity gain depends strongly on the propagation environment and diminishes with increasing correlation of the channel coefficients [13]. In higher frequency bands we expect an increase in correlation because the propagation channel becomes more and more line-of-sight (LOS) and we are confronted with a *rich array – poor scattering regime* [92].

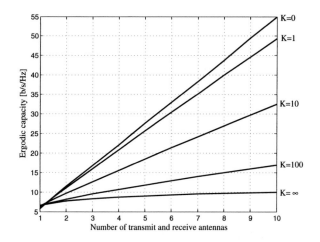

Fig. 6.1: Ergodic capacity of Ricean fading MIMO channels where K is the Ricean factor. For $K = 0$ we have the capacity for Rayleigh fading MIMO channels. For $K = \infty$ we have the capacity of the rank-1 LOS MIMO channel. `snr` was chosen to be 100, i.e., 20dB.

To illustrate the effect of a strong LOS component on the MIMO capacity we consider a $N \times N$ Ricean fading MIMO channel, which is given by [1]

$$\mathbf{H} = \sqrt{\frac{K}{K+1}}\overline{\mathbf{H}} + \sqrt{\frac{1}{K+1}}\mathbf{H}_{\mathrm{w}} \tag{6.1}$$

where $\overline{\mathbf{H}}$ is the fixed component and $\mathbf{H}_{\mathrm{w}}$ the fading component of the channel. For simplicity we assume for the fixed component $\overline{\mathbf{H}} = \mathbf{1}_N$ which is the all-one matrix, i.e., we neglect the phase differences between the channel gains[1]. For the fading component we assume a Rayleigh fading matrix, i.e., all entries of the matrix $\mathbf{H}_{\mathrm{w}}$ are i.i.d. Gaussian with zero mean and unit variance. The factor K is the Rice factor and denotes the ratio of the total power in the fixed component to the total power in the fading component. The ergodic capacity of the Ricean MIMO channel is then given by

$$C_{\mathrm{Rice}} = \mathcal{E}\left\{\log\left(\det\left(\mathbf{I}_N + \frac{\mathtt{snr}}{N}\mathbf{H}\mathbf{H}^{\mathrm{H}}\right)\right)\right\} \tag{6.2}$$

---

[1]However, in [119] it is shown that under certain geometrical assumptions about the distance between the transmitter and receiver and the antenna separations the phase differences are able to provide a full rank matrix in MIMO LOS channels. If the transmitter-receiver distance is much larger than the antenna separation at transmitter and receiver, it is a reasonable assumption to neglect the phase differences.

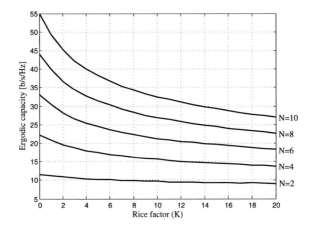

Fig. 6.2: Ergodic capacity vs. Ricean factor K for different numbers of transmit and receive antennas N.

where snr is the signal-to-noise ratio at each receive antenna and N the number of transmit antennas as well as receive antennas.

Fig. 6.1 shows the ergodic capacity as function of the number of antennas N for different Rice factors K. The highest capacity is achieved for $K = 0$, i.e., Rayleigh fading. For this case we can observe a linear increase with respect to the number of antennas. The smallest capacity is achieved for $K = \infty$, i.e., a nonfading MIMO channel. In that case the channel matrix has rank one and no spatial multiplexing gain is available. The logarithmic increase of the capacity is due to the receive array gain, i.e, the total SNR scales linearly with the number of receive antennas N. In Fig. 6.2 we plot the ergodic capacity as function of the Rice factor K for different number of antennas N. Again, we see the decrease of capacity with an increasing LOS component.

Other examples of rank-deficient or degenerated MIMO channels are MIMO channels with correlated fading [13] or keyhole channels [120]. In this chapter we will show that with two-hop relaying one can increase the rank and therefore the capacity of rank-deficient MIMO channels. We propose to use *amplify-and-forward relays* that act as active scatterers and assist the communication between a source terminal and destination a destination terminal: The relays receive the signal from the source in the first time slot and forward an amplified version to the destination in the second time slot. This way of relaying leads to low-complexity relay transceivers and to lower power consumption since there is no signal processing for decoding procedures. The goal of relaying here is to increase the rank of the compound (two time slots)

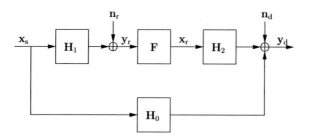

Fig. 6.3: System model for relay-assisted MIMO channel

channel matrix and to shape the eigenvalue distribution such that the channel matrix becomes well-conditioned and the capacity of the MIMO channel improves. We first look at the case when the relay antennas are connected through a wired backbone (distributed relay array) and the linear signal processing is done in a central unit. We then consider the case where the relays operate stand-alone, i.e, without a backbone connection.

6.2 Amplify-and-forward Protocols

We consider a multiple-input multiple-output (MIMO) system with N transmit antennas at the source terminal and N receive antennas at the destination terminal. The transmission of a data packet from the source to the destination terminal occupies two time slots of equal length. K single-antenna relay terminals receive signals during the first time slot and simultaneously forward a linearly modified version of the received signal during the second time slot. We assume that the relays cannot transmit and receive at the same time in the same frequency band (half-duplex mode). The variables of the system, cf. Fig.6.3, are:

- $\mathbf{x}_s$: $N \times 1$ transmit vector at the source with elements i.i.d. $\mathcal{CN}(0, P_s/N)$ and P_s the total power of the source terminal

- $\mathbf{y}_d$: $N \times 1$ receive vector at the destination

- $\mathbf{x}_r$: $K \times 1$ vector containing all transmit symbols at the K relays with $\mathcal{E}\left\{\mathbf{x}_r^H \mathbf{x}_r\right\} = P_r$ with P_r denoting the total average transmit power[2]

- $\mathbf{y}_r$: $K \times 1$ vector containing all received symbols at the K relays

---

[2] Averaging is done over source symbols and relay noise

- $\mathbf{H}_0$: $N \times N$ channel matrix between source and destination (direct channel)

- $\mathbf{H}_1$: $K \times N$ channel matrix between source and relays (first hop)

- $\mathbf{H}_2$: $N \times K$ channel matrix between relays and destination (second hop)

- $\mathbf{n}_d$: $N \times 1$ AWGN vector at the destination with elements i.i.d. $\mathcal{CN}(0, \sigma_d^2)$

- $\mathbf{n}_r$: $K \times 1$ vector containing AWGN of all relays with elements i.i.d. $\mathcal{CN}(0, \sigma_r^2)$

- $\mathbf{F}$: $K \times K$ forwarding matrix of the amplify-and-forward relays.

6.2.1 Signal and Channel Model

In time slot k the relay terminals receive the $K \times 1$ vector

$$\mathbf{y}_r[k] = \mathbf{H}_1[k]\mathbf{x}_s[k] + \mathbf{n}_r[k] \tag{6.3}$$

and the destination terminal receives the $N \times 1$ vector

$$\mathbf{y}_d[k] = \mathbf{H}_0[k]\mathbf{x}_s[k] + \mathbf{n}_d[k]. \tag{6.4}$$

In time slot $k + 1$ the destination terminal receives

$$\begin{aligned}
\mathbf{y}_d[k+1] &= \mathbf{H}_0[k+1]\mathbf{x}_s[k+1] + \mathbf{H}_2[k+1]\mathbf{x}_r[k+1] + \mathbf{n}_d[k+1] \\
&= \mathbf{H}_0[k+1]\mathbf{x}_s[k+1] + \mathbf{H}_2[k+1]\mathbf{F}[k]\mathbf{y}_r[k] + \mathbf{n}_d[k+1] \\
&= \mathbf{H}_0[k+1]\mathbf{x}_s[k+1] + \mathbf{H}_2[k+1]\mathbf{F}[k]\mathbf{H}_1[k]\mathbf{x}_1[k] \\
&\quad + \mathbf{H}_2[k+1]\mathbf{F}[k]\mathbf{n}_r[k] + \mathbf{n}_d[k+1].
\end{aligned} \tag{6.5}$$

The forwarding matrix $\mathbf{F} \in \mathcal{C}^{K \times K}$ maps the relay receive vector $\mathbf{y}_r$ to the relay transmit vector $\mathbf{x}_r = \mathbf{F}\mathbf{y}_r$. We will distinguish two different cases:

a) The relays cannot cooperate, i.e, they do not exchange channel or signal information. This model is appropriate if we assume that K *ad hoc relays* that operate stand-alone assist in the MIMO transmission between source and destination. The forwarding matrix $\mathbf{F}$ has a diagonal structure in this case.

b) The relays can fully and perfectly cooperate, i.e., each relay reports its received signal and its channel information to a central unit (without errors) where joint signal processing of all received signals can be established. This model is appropriate when the MIMO transmission between source and destination is assisted by a *distributed relay antenna system*. In this case the forwarding matrix $\mathbf{F}$ may have arbitrary structure.

From now on, we assume that the channel matrices $\mathbf{H}_0, \mathbf{H}_1, \mathbf{H}_2$ remain constant over at least two time slots and that $\{\mathbf{n}_r[k]\}_k$ and $\{\mathbf{n}_d[k]\}_k$ are stationary stochastic vector processes. Therefore, in the sequel we skip the time slot index k for the channels. Besides the assumption that we transmit over frequency-flat channels we do not impose in this section a specific model on the channel matrices. We stack the receive vectors $\mathbf{y}_d[k]$ and $\mathbf{y}_d[k+1]$ into one vector and obtain the following description of the two-hop MIMO relay channel:

$$
\begin{pmatrix} \mathbf{y}_d[k] \\ \mathbf{y}_d[k+1] \end{pmatrix} = \begin{bmatrix} \mathbf{H}_0 & \mathbf{0} \\ \mathbf{H}_2\mathbf{F}\mathbf{H}_1 & \mathbf{H}_0 \end{bmatrix} \begin{pmatrix} \mathbf{x}_s[k] \\ \mathbf{x}_s[k+1] \end{pmatrix} + \begin{bmatrix} 1 & 0 & 0 \\ 0 & 1 & \mathbf{H}_2\mathbf{F} \end{bmatrix} \begin{pmatrix} \mathbf{n}_d[k] \\ \mathbf{n}_d[k+1] \\ \mathbf{n}_r[k] \end{pmatrix} \qquad (6.6)
$$

based on this generic model, different transmission protocols can be specified:

Protocol P1: The source transmits $\mathbf{x}_s[k]$ in the first time slot to the destination and the relays. In the second time slot the relays forward the signal $\mathbf{x}_r[k+1]$ to the destination and the source transmits $\mathbf{x}_s[k+1]$ simultaneously to the destination. The signal structure is given in (6.6) and corresponds to the most general transmission protocol.

Protocol P2: The source transmits $\mathbf{x}_s[k]$ in the first time slot to the destination and the relays. In the second time slot the relays forward the signal $\mathbf{x}_r[k+1]$ to the destination whereas the source does not transmit in the second time slot. The signal structure simplifies to

$$
\begin{pmatrix} \mathbf{y}_d[k] \\ \mathbf{y}_d[k+1] \end{pmatrix} = \begin{bmatrix} \mathbf{H}_0 \\ \mathbf{H}_2\mathbf{F}\mathbf{H}_1 \end{bmatrix} \mathbf{x}_s[k] + \begin{bmatrix} 1 & 0 & 0 \\ 0 & 1 & \mathbf{H}_2\mathbf{F} \end{bmatrix} \begin{pmatrix} \mathbf{n}_d[k] \\ \mathbf{n}_d[k+1] \\ \mathbf{n}_r[k] \end{pmatrix} \qquad (6.7)
$$

Protocol P3. The source transmits $\mathbf{x}_s[k]$ in the first time slot only to the relays (the direct source-destination link is blocked for example due to shadowing). In the second time slot the relays forward the signal $\mathbf{x}_r[k+1]$ to the destination whereas the source does not transmit in the second time slot. The signal structure then follows as

$$
\mathbf{y}_d[k] = \mathbf{0} \qquad (6.8)
$$

$$
\mathbf{y}_d[k+1] = \mathbf{H}_2\mathbf{F}\mathbf{H}_1\mathbf{x}_s[k] + \mathbf{H}_2\mathbf{F}\mathbf{n}_r[k] + \mathbf{n}_d[k+1]. \qquad (6.9)
$$

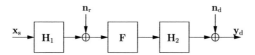

Fig. 6.4: System model for protocol P3

Another protocol proposed in [38] assumes that the source transmits in every time slot but the destination does not observe the source signal in the first time slot (for example because the destination is busy due to other tasks). In [41], [38] it was shown that protocol P1 performs best in terms of achievable rate, see also Section 6.2.5. However, we will assume that the direct channel $\mathbf{H}_0$ is rank-deficient and does not provide enough spatial degrees of freedom to provide multiplexing gain. Including the observation from the direct channel by combining the signals from the first and the second time slot (for example by maximum ratio combining) leads to an SNR gain. Since we are interested in how AF relays can increase the spatial multiplexing capability we focus on protocol P3, illustrated in Fig.6.4.

Let

$$\mathbf{H}_1 = \mathbf{U}_1 \boldsymbol{\Sigma}_1 \mathbf{V}_1^H \tag{6.10}$$

$$\mathbf{H}_2 = \mathbf{U}_2 \boldsymbol{\Sigma}_2 \mathbf{V}_2^H \tag{6.11}$$

be the singular value decompositions (SVD) of the first-hop and second-hop channel matrices, respectively. The receive vector $\mathbf{y}_d$ is then rewritten as[3]

$$\mathbf{y}_d = \mathbf{U}_2 \boldsymbol{\Sigma}_2 \mathbf{V}_2^H \mathbf{F} \mathbf{U}_1 \boldsymbol{\Sigma}_1 \mathbf{V}_1^H \mathbf{x}_s + \mathbf{U}_2 \boldsymbol{\Sigma}_2 \mathbf{V}_2^H \mathbf{F} \mathbf{n}_r + \mathbf{n}_d. \tag{6.12}$$

For $\widetilde{\mathbf{x}}_s = \mathbf{V}_1^H \mathbf{x}_s$ we have $\mathcal{E}\left\{\widetilde{\mathbf{x}}_s \widetilde{\mathbf{x}}_s^H\right\} = \mathcal{E}\left\{\mathbf{x}_s \mathbf{x}_s^H\right\} = P_s/N\, \mathbf{I}_N$, i.e., multiplication of $\mathbf{x}_s$ with the unitary matrix $\mathbf{V}_1^H$ does not change the covariance matrix of $\mathbf{x}_s$. It follows for the mutual information $I\left(\widetilde{\mathbf{x}}_s; \mathbf{y}_d\right) = I(\mathbf{x}_s; \mathbf{y}_d)$. For the same reasons we may multiply (6.9) at the destination terminal with the unitary matrix $\mathbf{U}_2^H$ to obtain an equivalent system model (see Figs. 6.4 and 6.5):

$$\widetilde{\mathbf{y}}_d = \boldsymbol{\Sigma}_2 \mathbf{V}_2^H \mathbf{F} \mathbf{U}_1 \boldsymbol{\Sigma}_1 \widetilde{\mathbf{x}}_s + \boldsymbol{\Sigma}_2 \mathbf{V}_2^H \mathbf{F} \mathbf{n}_r + \widetilde{\mathbf{n}}_d, \tag{6.13}$$

where $\widetilde{\mathbf{y}}_d = \mathbf{U}_2^H \mathbf{y}_d$ and $\widetilde{\mathbf{n}}_d = \mathbf{U}_2^H \mathbf{n}_d$. Note that $I(\widetilde{\mathbf{x}}_s; \widetilde{\mathbf{y}}_d) = I\left(\widetilde{\mathbf{x}}_s; \mathbf{y}_d\right) = I(\mathbf{x}_s; \mathbf{y}_d)$. In the following we will drop the tilde notation for the equivalent signals $\widetilde{\mathbf{x}}_s$, $\widetilde{\mathbf{y}}_d$ and $\widetilde{\mathbf{n}}_d$ and use $\mathbf{x}_s$, $\mathbf{y}_d$

---

[3]We drop the time index k completely.

151

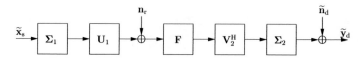

Fig. 6.5: Equivalent system model for protocol P3

and $\mathbf{n}_d$ instead. The equivalent system model (6.13) is useful when we assume complete channel state information at the relays for the design of the forwarding matrix $\mathbf{F}$.

6.2.2 Achievable Rates

We assume that the elements of $\mathbf{H}_1$ and $\mathbf{H}_2$ are frequency-flat and time varying. We use a block-fading channel model [121] where a channel gain remains constant during a certain time interval (coherence interval) and changes independently from interval to interval. We refer to *slow fading* when the coherence interval is too long for a codeword to span several coherence intervals. In this case the ϵ-outage capacity [122] is an appropriate performance measure. We speak of *fast fading* when the codeword length spans several coherence intervals of the channel. That might be due to less stringent delay constraints or due to a fast varying channel. In this situation the ergodic capacity [122] is the appropriate performance measure. In this chapter we focus on ergodic capacity, since spatial multiplexing gains ar usually expressed in terms of ergodic capacity.

The mutual information in bits/s/Hz for perfectly known $\mathbf{H}_1$ and $\mathbf{H}_2$ at the destination terminal is given by

$$I(\mathbf{x}_s; \mathbf{y}_d) = \frac{1}{2} \log_2 \det \left(\mathbf{I}_N + \frac{P_s}{N} \mathbf{R}^{-1}(\mathbf{F}) \mathbf{H}(\mathbf{F}) \mathbf{H}^H(\mathbf{F}) \right) \tag{6.14}$$

with the equivalent $N \times N$ MIMO channel (over two hops)

$$\mathbf{H}(\mathbf{F}) = \mathbf{\Sigma}_2 \mathbf{V}_2^H \mathbf{F} \mathbf{U}_1 \mathbf{\Sigma}_1 \tag{6.15}$$

and the noise covariance matrix

$$\mathbf{R}(\mathbf{F}) = \mathcal{E} \left\{ \left(\mathbf{\Sigma}_2 \mathbf{V}_2^H \mathbf{F} \mathbf{n}_r + \mathbf{n}_d \right) \left(\mathbf{\Sigma}_2 \mathbf{V}_2^H \mathbf{F} \mathbf{n}_r + \mathbf{n}_d \right)^H \right\} \tag{6.16}$$

$$= \sigma_r^2 \mathbf{\Sigma}_2 \mathbf{V}_2^H \mathbf{F} \mathbf{F}^H \mathbf{V}_2 \mathbf{\Sigma}_2^H + \sigma_d^2 \mathbf{I}_N. \tag{6.17}$$

Reliable communication from source to destination terminal is possible whenever

$$\frac{1}{2} \log_2 \det \left(\mathbf{I}_N + \frac{P_s}{N} \mathbf{R}^{-1}(\mathbf{F}) \mathbf{H}(\mathbf{F}) \mathbf{H}^{\mathrm{H}}(\mathbf{F}) \right) > R, \tag{6.18}$$

where R is the target rate in bits/s/Hz. When the equivalent MIMO channel does not satisfy condition (6.18) the system is in outage. The probability of an outage event is given by:

$$p_{\mathrm{out}}(R, \mathbf{F}) = \mathbb{P}\left[I(\mathbf{x}_s; \mathbf{y}_d) < R \right] \tag{6.19}$$

$$= \mathbb{P}\left[\frac{1}{2} \log_2 \det \left(\mathbf{I}_N + \frac{P_s}{N} \mathbf{R}^{-1}(\mathbf{F}) \mathbf{H}(\mathbf{F}) \mathbf{H}^{\mathrm{H}}(\mathbf{F}) \right) < R \right]. \tag{6.20}$$

The ϵ-outage capacity $C_{\epsilon,\mathbf{F}}$ is obtained by solving $p_{\mathrm{out}}(R = C_\epsilon, \mathbf{F}) = \epsilon$. By coding over a large number of coherence intervals of the channel, the ergodic capacity serves as a long-term rate of reliable communication:

$$C_{\mathrm{erg}}(\mathbf{F}) = \mathcal{E}\left\{ I(\mathbf{x}_s; \mathbf{y}_d) \right\} \tag{6.21}$$

$$= \mathcal{E}\left\{ \frac{1}{2} \log_2 \det \left(\mathbf{I}_N + \frac{P_s}{N} \mathbf{R}^{-1}(\mathbf{F}) \mathbf{H}(\mathbf{F}) \mathbf{H}^{\mathrm{H}}(\mathbf{F}) \right) \right\}. \tag{6.22}$$

Depending on the degree of channel state information available at the relay terminals, different possibilities for the optimum choice of the forwarding matrix exist. For example for average CSI at the relay terminals one might choose:

$$C_{\mathrm{erg}} = \max_{\mathbf{F} \in \mathcal{F}} \mathcal{E}\left\{ \frac{1}{2} \log_2 \det \left(\mathbf{I}_N + \frac{P_s}{N} \mathbf{R}^{-1}(\mathbf{F}) \mathbf{H}(\mathbf{F}) \mathbf{H}^{\mathrm{H}}(\mathbf{F}) \right) \right\} \tag{6.23}$$

where $\mathcal{F}$ is the set of all forwarding matrices that fulfil the power constraint of the relays. When each channel realization is known the optimization can be written as:

$$C_{\mathrm{erg}} = \mathcal{E}\left\{ \max_{\mathbf{F} \in \mathcal{F}} \frac{1}{2} \log_2 \det \left(\mathbf{I}_N + \frac{P_s}{N} \mathbf{R}^{-1}(\mathbf{F}) \mathbf{H}(\mathbf{F}) \mathbf{H}^{\mathrm{H}}(\mathbf{F}) \right) \right\}. \tag{6.24}$$

Different solutions arise depending on whether the relays know both channels $\mathbf{H}_1$ and $\mathbf{H}_2$ or only $\mathbf{H}_1$.

In the following we first consider the case of a *distributed relay array* (DRA) where the relay terminals are connected through a wired backbone and the signals can be jointly processed at a central processor attached to the backbone. Then, we look at the case of *ad hoc relays* (AR) where the relay terminals do not cooperate with each other, i.e., each relay terminal only processes its own receive signal and forwards it to the destination. In this situation the forwarding matrix $\mathbf{F}$ is diagonal.

6.2.3 System with Distributed Relay Array (DRA)

If the relays are connected through a wired backbone and the forwarding matrix can have arbitrary structure, since the receive signal at each relay is known to all relays. We assume that the wired backbone is perfect, i.e., the receive signal (including noise) at each relay is perfectly known at a central processor. We first look at the case, where the relay has perfect knowledge of each channel realization of $\mathbf{H}_1$ and $\mathbf{H}_2$. Then we consider the case where the relay has only knowledge of $\mathbf{H}_1$.

Case 1: We assume that the destination terminal and the relay terminals have perfect knowledge of $\mathbf{H}_1$ and $\mathbf{H}_2$ and the source terminal has only knowledge about the channel distributions of $\mathbf{H}_1$ and $\mathbf{H}_2$. Furthermore, the signals received at the relay terminals may be processed jointly at a central processor connected to the backbone of the distributed relay array. Without loss of generality we may expand the forwarding matrix $\mathbf{F}$ as [91]

$$\mathbf{F} = \mathbf{V}_2 \mathbf{F}' \mathbf{U}_1^{\mathrm{H}} \tag{6.25}$$

where $\mathbf{V}_2$ and $\mathbf{U}_1^{\mathrm{H}}$ are taken from (6.11) and (6.10). $\mathbf{F}' \in \mathbb{R}^{K \times K}$ is called the *inner forwarding matrix*. The receive signal at the destination terminal is then given as

$$\mathbf{y}_{\mathrm{d}} = \mathbf{\Sigma}_2 \mathbf{F}' \mathbf{\Sigma}_1 \mathbf{x}_{\mathrm{s}} + \mathbf{\Sigma}_2 \mathbf{F}' \widetilde{\mathbf{n}}_{\mathrm{r}} + \mathbf{n}_{\mathrm{d}}, \tag{6.26}$$

where $\widetilde{\mathbf{n}}_{\mathrm{r}} = \mathbf{U}_1^{\mathrm{H}} \mathbf{n}_{\mathrm{r}}$. Note that (6.25) can be interpreted as distributed receive beamforming (with $\mathbf{U}_1^{\mathrm{H}}$) in the first time slot and distributed transmit beamforming (with $\mathbf{V}_2$) in the second time slot. The equivalent $N \times N$ MIMO channel (over two hops) follows as

$$\mathbf{H}(\mathbf{F}') = \mathbf{\Sigma}_2 \mathbf{F}' \mathbf{\Sigma}_1 \tag{6.27}$$

and the noise covariance matrix at the destination terminal as

$$\mathbf{R}(\mathbf{F}') = \sigma_{\mathrm{r}}^2 \mathbf{\Sigma}_2 \mathbf{F}'^2 \mathbf{\Sigma}_2^{\mathrm{H}} + \sigma_{\mathrm{d}}^2 \mathbf{I}_N. \tag{6.28}$$

The equivalent system model is given in Fig.6.6 for the case $K = N$, where μ_i is the ith singular value of the first-hop channel $\mathbf{H}_1$ and λ_i the ith singular value of the second-hop channel $\mathbf{H}_2$.

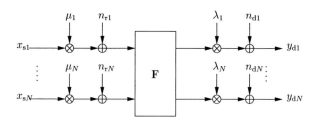

Fig. 6.6: $K = N$

The mutual information between $\mathbf{x}_s$ and $\mathbf{y}_d$ is given by

$$I(\mathbf{x}_s; \mathbf{y}_d) = \frac{1}{2} \log_2 \det \left(\mathbf{I}_N + \frac{\text{snr}}{N} \left(\Sigma_2 \mathbf{F}'^2 \Sigma_2 + \mathbf{I}_N \right)^{-1} \Sigma_2 \mathbf{F}' \Sigma_1^2 \mathbf{F}'^T \Sigma_2 \right) \qquad (6.29)$$

where $\text{snr} = P/\sigma^2$, $P = P_s = P_r$ and $\sigma^2 = \sigma_d^2 = \sigma_r^2$ for simplicity. When we choose the inner forwarding matrix to be diagonal it has been shown in [123] that the optimal diagonal elements are given by

$$|f_i'|^2 = \frac{1}{\frac{P}{N} \mu_i + \sigma^2} \cdot p_i \qquad (6.30)$$

for $i = 1, 2, \ldots, N$ with

$$p_i = \left[0, \sqrt{\frac{P\mu_i}{\eta N \lambda_i^2} + \left(\frac{P\mu_i^2}{2\eta N \lambda_i^2} \right)^2} - \frac{P\mu_i^2}{2\eta N \lambda_i^2} - \frac{\sigma^2}{\lambda_i^2} \right]^+ \qquad (6.31)$$

where $[0, x]^+ = \max(0, x)$ and η is the waterfilling level which is chosen such, that the total relay power constraint is met.

Case 2: In the following, we consider the case that the central unit knows the first hop channel $\mathbf{H}_1$ perfectly, but has no CSI about the second-hop channel $\mathbf{H}_2$. Furthermore, we assume that the second-hop channel has no preferred direction, i.e., the elements of the channel matrix are zero mean and i.i.d. distributed. For this scenario, the optimal forwarding matrix

$$\mathbf{F}^{\text{opt}} = \arg \max_{\mathbf{F} \in \mathcal{F}} \mathcal{E}_{\mathbf{H}_2} \left(\log \det(\mathbf{I}_N + \frac{\text{snr}}{N} (\mathbf{I}_N + \mathbf{H}_2 \mathbf{F} \mathbf{F}^H \mathbf{H}_2^H)^{-1} \mathbf{H}_2 \mathbf{F} \mathbf{H}_1 \mathbf{H}_1^H \mathbf{F}^H \mathbf{H}_2^H) \right) \qquad (6.32)$$

subject to

$$\text{tr} \left(\frac{\text{snr}}{N} \mathbf{H}_1 \mathbf{F} \mathbf{F}^H \mathbf{H}_1^H + \mathbf{F} \mathbf{F}^H \right) = \text{snr} \qquad (6.33)$$

is not known in the general case. However, some insight into the structure of the optimal matrix can be obtained by considering the asymptotic regimes $\mathtt{snr} \to 0$, $\mathtt{snr} \to \infty$ and $N \to \infty$. Firstly, we investigate the low SNR case. Here, it can be shown that choosing the forwarding matrix as a matched filter with respect to the channel of the first hop, i.e.,

$$\mathbf{F} = \sqrt{\frac{\mathtt{snr}}{(\mathtt{snr} \cdot \mathrm{tr}(\mathbf{H}_1 \mathbf{H}_1^{\mathrm{H}}) + 1)}} \cdot \frac{\mathbf{H}_1^{\mathrm{H}}}{\sqrt{\mathrm{tr}(\mathbf{H}_1 \mathbf{H}_1^{\mathrm{H}})}}, \tag{6.34}$$

is asymptotically optimal. In order to show this, we approximate the objective function in (6.32) as follows:

$$\mathcal{E}_{\mathbf{H}_2}\left\{I(\mathbf{x}_{\mathrm{s}}; \mathbf{y}_{\mathrm{d}})\right\} \approx \mathcal{E}_{\mathbf{H}_2}\left\{\log \det(\mathbf{I}_N + \frac{\mathtt{snr}}{N} \cdot \mathbf{H}_2 \mathbf{F} \mathbf{H}_1 \mathbf{H}_1^{\mathrm{H}} \mathbf{F}^{\mathrm{H}} \mathbf{H}_2^{\mathrm{H}})\right\} \tag{6.35}$$

$$\approx \log(e) \cdot \frac{\mathtt{snr}}{N} \cdot \mathcal{E}_{\mathbf{H}_2}\left\{\mathrm{tr}(\mathbf{H}_2 \mathbf{F} \mathbf{H}_1 \mathbf{H}_1^{\mathrm{H}} \mathbf{F}^{\mathrm{H}} \mathbf{H}_2^{\mathrm{H}})\right\} \tag{6.36}$$

$$= \log(e) \cdot \frac{\mathtt{snr}}{N} \cdot \mathrm{tr}(\mathbf{H}_1 \mathbf{F} \mathcal{E}\left\{\mathbf{H}_2 \mathbf{H}_2^{\mathrm{H}}\right\} \mathbf{F}^{\mathrm{H}} \mathbf{H}_1^{\mathrm{H}}) \tag{6.37}$$

$$= \log(e) \cdot \mathtt{snr} \cdot \mathrm{tr}(\mathbf{F} \mathbf{H}_1 \mathbf{H}_1^{\mathrm{H}} \mathbf{F}^{\mathrm{H}}). \tag{6.38}$$

The first approximation follows from $\lim_{\mathtt{snr} \to 0}(\mathbf{I}_N + \mathbf{H}_2 \mathbf{F} \mathbf{F}^{\mathrm{H}} \mathbf{H}_2^{\mathrm{H}}) = \mathbf{I}_N$, the second one is the well known MIMO low SNR approximation of mutual information, we may swap expectation and trace due to linearity and the rearrangement of $\mathbf{H}_1$ and $\mathbf{H}_2$ follows as matrices commute under multiplication inside the trace. Thus, it remains to solve the following problem, whose solution in (6.34) follows immediately from the fact that matched filtering maximizes SNR:

$$\mathbf{F}_{\mathrm{low\ snr}}^{\mathrm{opt}} = \arg \max_{\mathbf{F}:\mathrm{tr}(\mathbf{F}\mathbf{F}^{\mathrm{H}})=\mathtt{snr}} \mathrm{tr}(\mathbf{F} \mathbf{H}_1 \mathbf{H}_1^{\mathrm{H}} \mathbf{F}^{\mathrm{H}}). \tag{6.39}$$

This solution is also intuitive as spatial multiplexing gain vanishes in the low SNR regime, and thus SNR maximization becomes crucial for rate maximization.

In the high SNR regime, again, we use an appropriate approximation which allows for the identification of the important design criteria for the forwarding matrix:

$$\mathcal{E}_{\mathbf{H}_2}\left\{I(\mathbf{x}_{\mathrm{s}}; \mathbf{y}_{\mathrm{ds}})\right\} \approx N \log \mathtt{snr} + \log \det(\mathbf{F} \mathbf{H}_1 \mathbf{H}_1^{\mathrm{H}} \mathbf{F}^{\mathrm{H}}) \tag{6.40}$$

$$+ \mathcal{E}_{\mathbf{H}_2}\left\{\log \det(\mathbf{H}_2 \mathbf{H}_2^{\mathrm{H}})\right\} \tag{6.41}$$

$$- \mathcal{E}_{\mathbf{H}_2}\left\{\log \det(\mathbf{I}_N + \mathbf{H}_2 \mathbf{F} \mathbf{F}^{\mathrm{H}} \mathbf{H}_2^{\mathrm{H}})\right\}. \tag{6.42}$$

The optimization problem thus reduces to

$$\mathbf{F}_{\text{high snr}}^{\text{opt}} = \arg \max_{\mathbf{F} \in \mathcal{F}} \log \det(\mathbf{F} \mathbf{H}_1 \mathbf{H}_1^{\text{H}} \mathbf{F}^{\text{H}}) - \mathcal{E}_{\mathbf{H}_2} \left\{ \log \det(\mathbf{I}_N + \mathbf{H}_2 \mathbf{F} \mathbf{F}^{\text{H}} \mathbf{H}_2^{\text{H}}) \right\} \qquad (6.43)$$

such that

$$\text{tr} \left(\frac{\text{snr}}{N} \mathbf{H}_1 \mathbf{F} \mathbf{F}^{\text{H}} \mathbf{H}_1^{\text{H}} \right) = \text{snr}. \qquad (6.44)$$

Although we are not aware of an analytic solution, both terms can be identified to reflect two conflicting requirements. By applying the arithmetic mean-geometric mean inequality, the first term is found to be maximized by a zero-forcing (ZF) matrix $\mathbf{F} \propto (\mathbf{H}_1^{\text{H}} \mathbf{H}_1)^{-1} \mathbf{H}_1^{\text{H}}$. Consequently, it accentuates the need for isotropic signal radiation into the second hop channel. The second term on the other hand is nothing else but the ergodic mutual information $\mathcal{E} \{ I(\mathbf{n}_r; \mathbf{y}_d) \}$ evaluated at a low SNR which is mainly kept small by avoiding noise enhancement at the relay. In conclusion, the optimal forwarding matrix thus has to balance isotropic radiation with noise enhancement. A filter well known to accomplish a closely related tradeoff, namely the one between multi-stream interference and noise enhancement, is realized by the minimum-mean-squared-error (MMSE) matrix. Such a forwarding matrix demonstrates better performance than the ZF matrix in our computer experiments, and can be expected to perform close to optimal.

6.2.4 System with Ad Hoc Relays

Here we assume that the relays are not connected through a wired backbone, and therefore are not aware of any other received signals than their own ones. Thus, the forwarding matrix is constrained to be diagonal. We will assume that $N \gg 1$ in this section, such that the law of large numbers becomes effective and each relay receives roughly the same signal power. The optimal forwarding matrix – due to symmetry – then is $\mathbf{F} \propto \mathbf{I}_K$.

We are interested in the performance of this particularly simple choice of $\mathbf{F}$ compared to the forwarding strategies introduced in the previous section. While it is clearly inferior to the matched filter in the low SNR regime as it does not collect the signal power efficiently, the result in the medium and high SNR region looks different. Denoting the rate under $\mathbf{F} \propto \mathbf{I}_K$ by $R_\mathbf{I}$ and the rate under an MMSE forwarding matrix[4](for the case of a distributed relay array with CSI of thew first hop) by R_{MMSE} we have the following:

Given a certain SNR there is always an N_0, such that for $N > N_0$ we have $R_\mathbf{I} > R_{\text{MMSE}}$.

---

[4]The MMSE forwarding matrix is given by $\mathbf{F}_{\text{MMSE}}$

To gain insight into the reason behind this, we let N grow large and keep K fixed, such that

$$\frac{1}{N} \cdot \mathbf{H}_i \mathbf{H}_i^H \longrightarrow \mathbf{I}_N \text{ as } N \to \infty, \text{ for } i = 1, 2 \tag{6.45}$$

again by the law of large numbers. Thus, choosing $\mathbf{F} = \alpha \mathbf{I}_K$ (the same derivations hold if $\mathbf{F}$ is unitary) yields

$$\mathbf{H} = \frac{1}{N} \mathbf{H}_2 \mathbf{F} \mathbf{H}_1 \mathbf{H}_1^H \mathbf{F}^H \mathbf{H}_2^H \longrightarrow \mathbf{H}_2 \mathbf{F} \mathbf{F}^H \mathbf{H}_2^H = \alpha \mathbf{H}_2 \mathbf{H}_2^H \tag{6.46}$$

and

$$\mathbf{R} = \frac{1}{N} \mathbf{H}_2 \mathbf{F} \mathbf{F}^H \mathbf{H}_2^H + \mathbf{I}_N = \frac{\alpha}{N} \mathbf{H}_2 \mathbf{H}_2^H + \mathbf{I}_N \longrightarrow (1 + \alpha) \mathbf{I}_N. \tag{6.47}$$

Consequently, this choice of $\mathbf{F}$ yields an (almost) isotropic radiation into the second hop channel, while – in contrast to the MMSE matrix – avoiding both noise enhancement and coloring. Thus, it will demonstrate better performance for sufficiently large N, if $\mathbf{H}_2$ has no preferred direction. Our computer simulations even show that for moderate SNR we have $N_0 = 1$, although convergence in (6.45) is rather slow. The loss in performance compared to the distributed relay array with knowledge of the first hop channel at the relay only can thus be expected to be small for sufficiently large N under moderate and high SNR. This result is also appealing for the linear distributed array system scenario due to its simplicity. In Section 6.3 we will look at the ad hoc relay case in more detail for a specific channel model and show how ad hoc AF relays can help to provide spatial multiplexing gain in poor scattering environments.

6.2.5 Numerical Examples

In this section we give some numerical examples for Gaussian i.i.d. fading channels. Fig.6.7 compares the ergodic capacities of protocols P1–P3 for a $4 \times 4 \times 4$ system, i.e. 4 antennas at source and destination and 4 relays. We can see that protocol P1 performs best in terms of ergodic capacity. We normalized the transmit powers such that each node consumes an average transmit energy of PT over two time slots of length T. SNR is defined as P/σ^2, where σ^2 is the noise variance at each antenna (relay or destination). Fig.6.8 shows the CDFs of the mutual information of protocols P1–P3. We observe that protocols P2 and P3 have slightly higher diversity orders ("steeper" CDF) than protocol P1. The reason is that in protocol P1 every second transmit vector interferes with the transmit vector of the relays. Since both vectors do not contain the same data, the different degrees of freedom have to be used for separation of the signals rather than combining them to achieve higher diversity gains.

We also confirm some of the results obtained in the previous section through numerical

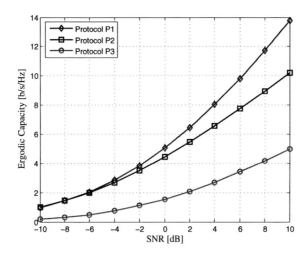

Fig. 6.7: Ergodic capacities for protocols P1–P3 for $N = K = 4$ and ad hoc relays

examples. In Fig. 6.9 and 6.10 we plot the capacity of the relay channel under different forwarding matrices. While $\mathbf{F} \propto \mathbf{I}$ can be realized in both linear distributed array and ad hoc relay systems, all other strategies are applicable to the first case only.

As a reference, the capacity in the case that the relay array knows both $\mathbf{H}_1$ and $\mathbf{H}_2$ and chooses the forwarding matrix in an optimal way (6.30) is plotted.

Under low SNR, we see that the (asymptotically optimal) MF matrix performs best among the matrices realizable without channel knowledge about $\mathbf{H}_2$ and approaches the performance of the MMSE matrix from above. $\mathbf{F} \propto \mathbf{I}$ performs poor as it does not collect the signal power efficiently, the ZF matrix even worse as it enhances the noise. Under high SNR the MMSE matrix confirms to balance noise enhancement and isotropic radiation into the second hop and shows best performance as expected. The ZF matrix ensures perfect isotropic radiation, however, the noise enhancement – though not crucial in this case – makes it inferior to the MMSE matrix. We can also see that indeed the identity matrix becomes the best choice among the considered candidates under medium SNR. The intersection point of its curve with the curve of the MMSE matrix in Fig. 6.9 will be shifted even further to the right, if N is increased. This leads to the conclusion that whenever the second-hop channel is not known at the relay array the joint processing of the received signals at the relay antennas does not lead to significant performance gains for sufficiently high SNR. Using a diagonal forwarding matrix is sufficient which means that the

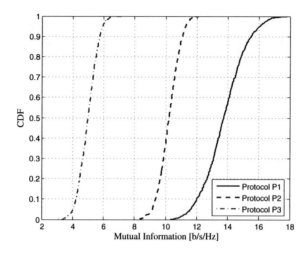

Fig. 6.8: CDFs of mutual information of protocols P1–P3 for $N = K = 4$ and ad hoc relays

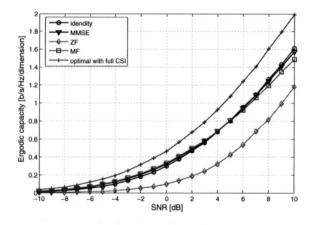

Fig. 6.9: Ergodic capacities for low SNR and different choices of $\mathbf{F}$ and $N = K = 4$

relay antennas have not to be connected to a central unit. An intuitive argument strengthening the derivations in the large array limit in the previous section is that unitary matrices do not diminish the rank of the compound channel matrix, while simultaneously keeping the noise level constant.

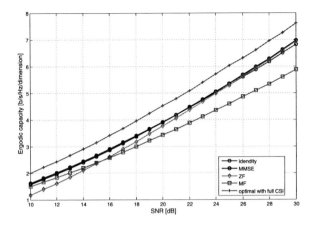

Fig. 6.10: Ergodic capacities for high SNR and different choices of **F**

6.3 Ergodic Performance of a Relay-assisted Rank-deficient MIMO Channel

In this section we assume that the relays are not connected to a central unit (ad hoc relays) and we assume a specific channel model for the first-hop and second-hop matrix channels. We assume that the channel gain from one source antenna to one relay antenna[5] is characterized by an angle of departure (AoD) and a path gain that is a complex Gaussian random variable. The channel gain from the relay antenna to one destination antenna is characterized by an angle of arrival (AoA) and again a path gain that is again a complex Gaussian random variables. We assume that the relays do not move during the time of interest and that the path gains of the first and the second hop channels are independent. Further, we look at the case where the relay operate stand-alone, i.e., they are not connected to a central processor where joint signal processing can be done. Hence, the relay can only operate in an amplify-and-forward mode since single-antenna decode-and-forward relays cannot decode the MIMO encoded source signal.

---

[5]Each relay is equipped with only one antenna.

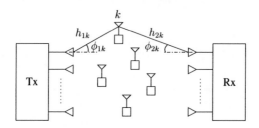

Fig. 6.11: Relay-assisted MIMO communication system

6.3.1 System Model

We consider uniform linear antenna arrays at the source and the destination and single-antenna relays. For the antenna array depicted in Fig. 6.11 we denote the AoD-vector with respect to relay k as

$$\boldsymbol{\Phi}_{1k} = \left[1, e^{j2\pi d \sin \phi_{1k}/\lambda}, \ldots, e^{j2\pi(N-1)d \sin \phi_{1k}/\lambda}\right]^{\mathrm{T}}, \tag{6.48}$$

where d is the antenna spacing at source and destination, λ the operational wavelength and $\phi_{1k}, \in [-\pi/2, \pi/2]$ the horizontal angle characterizing the paths to relay k. (6.48) follows from a narrowband signal and planar wavefront assumption [124]. The array response for a plane wave arriving at angle ϕ_{2k} follows accordingly as

$$\boldsymbol{\Phi}_{2k} = \left[1, e^{-j2\pi d \sin \phi_{2k}/\lambda}, \ldots, e^{-j2\pi(N-1)d \sin \phi_{2k}/\lambda}\right]^{\mathrm{T}}, \tag{6.49}$$

where $\phi_{2k} \in [-\pi/2, \pi/2]$ are the horizontal angles characterizing the paths from relay k, respectively (Fig. 6.11). The signal received in the first time slot by relay k is

$$y_{\mathrm{r},k} = h_{1k} \boldsymbol{\Phi}_{1k}^{\mathrm{T}} \mathbf{x}_{\mathrm{s}} + n_{\mathrm{r}k} \tag{6.50}$$

where h_{1k} the first hop channel coefficient of relay k which usually accounts for path loss, shadowing and small-scale fading. We assume that the fading coefficient is approximately the same for all transmit antennas (spacing of antenna elements at both sides sufficiently small), i.e.,

$$h_{1k} = h_{1k}^{(1)} \approx h_{1k}^{(2)} \approx \ldots \approx h_{1k}^{(N)}, \tag{6.51}$$

where $h_{1k}^{(j)}$ denotes the channel coefficient between transmit antenna j and relay antenna k. Note that we consider protocol P1 and therefore the destination does not receive any signal from the source in the first time slot. In the second time slot the destination receives

$$\mathbf{y}_d = \sum_{k=1}^{K} h_{2k} f_k h_{1k} \mathbf{\Phi}_{2k} \mathbf{\Phi}_{1k}^T \mathbf{x}_s + \sum_{k=1}^{K} h_{2k} f_k \mathbf{\Phi}_{2k} n_{rk} + \mathbf{n}_d, \qquad (6.52)$$

where h_{2k} is the second hop channel coefficient of relay k (again approximately the same for all receive antennas), f_k is the gain factor of relay k and $\mathbf{y}_d = [y_{d,1}, \ldots, y_{d,N}]^T$ is the receive vector at the destination in time slot 2. In matrix notation we write

$$\mathbf{y}_d = \mathbf{\Phi}_2 \mathbf{\Gamma}_{12} \mathbf{\Phi}_1^T \mathbf{x}_s + \mathbf{\Phi}_2 \mathbf{\Gamma}_2 \mathbf{n}_r + \mathbf{n}_d, \qquad (6.53)$$

where $\mathbf{\Gamma}_{12}$ and $\mathbf{\Gamma}_2$ are diagonal with $[\mathbf{\Gamma}_{12}]_{k,k} = h_{2k} f_k h_{1k}$ and $[\mathbf{\Gamma}_2]_{k,k} = h_{2k} f_k$. The columns of $\mathbf{\Phi}_1$ are the steering vectors (6.48) and the columns of $\mathbf{\Phi}_2$ are the array response vectors (6.49) and $\mathbf{n}_3 = [n_{31}, \ldots, n_{rK}]$.

6.3.2 Achievable Rate

The mutual information of (6.53) is given by

$$I\left(\mathbf{x}_s; \mathbf{y}_d | \mathbf{\Phi}_2 \mathbf{\Gamma}_{12} \mathbf{\Phi}_1^T, \mathbf{R}\right) = h\left(\mathbf{y}_d | \mathbf{\Phi}_2 \mathbf{\Gamma}_{12} \mathbf{\Phi}_1^T, \mathbf{R}\right) - h\left(\mathbf{n} | \mathbf{\Phi}_2 \mathbf{\Gamma}_{12} \mathbf{\Phi}_1^T, \mathbf{R}\right) \qquad (6.54)$$

where $h(\cdot)$ denotes the differential entropy of a random vector, $\mathbf{n} = \mathbf{\Phi}_2 \mathbf{\Gamma}_2 \mathbf{n}_r + \mathbf{n}_d$ and $\mathbf{R}$ is the covariance matrix of the effective noise at the destination

$$\mathbf{R} = \mathbf{\Phi}_2 \mathbf{\Gamma}_{12} \mathbf{\Gamma}_{12}^H \mathbf{\Phi}_2^H \sigma_r^2 + \sigma_d^2 \mathbf{I}_N. \qquad (6.55)$$

The mutual information is maximized when $h\left(\mathbf{y}_d | \mathbf{\Phi}_2 \mathbf{\Gamma}_{12} \mathbf{\Phi}_1^T, \mathbf{R}\right)$ is maximized, i.e., the receive vector $\mathbf{y}_d$ has to be Gaussian for a given two-hop channel matrix $\mathbf{\Phi}_2 \mathbf{\Gamma}_{12} \mathbf{\Phi}_1^T$ and its differential entropy is [74]

$$h\left(\mathbf{y}_d | \mathbf{\Phi}_2 \mathbf{\Gamma}_{12} \mathbf{\Phi}_1^T, \mathbf{R}\right) = \log_2 \det\left(\pi e \left(\mathbf{\Phi}_2 \mathbf{\Gamma}_{12} \mathbf{\Phi}_1^T \mathbf{R}_s \mathbf{\Phi}_1^* \mathbf{\Gamma}_{12}^H \mathbf{\Phi}_2^H\right) + \mathbf{R}\right), \qquad (6.56)$$

where $\mathbf{R}_s = \mathrm{E}\left\{\mathbf{x}_s \mathbf{x}_s^H\right\}$. The mutual information measured in bits per channel use follows then as

$$I\left(\mathbf{x}_s; \mathbf{y}_d | \mathbf{\Phi}_2 \mathbf{\Gamma}_{12} \mathbf{\Phi}_1^T, \mathbf{R}\right) = \frac{1}{2} \sum_{k=1}^{r} \log_2\left(1 + \frac{P_s}{N} \lambda_k\left(\mathbf{R}^{-1} \mathbf{\Phi}_2 \mathbf{\Gamma}_{12} \mathbf{\Phi}_1^T \mathbf{\Phi}_1^* \mathbf{\Gamma}_{12}^H \mathbf{\Phi}_2^H\right)\right), \qquad (6.57)$$

where

$$r = \mathrm{rk}\left(\mathbf{R}^{-1}\boldsymbol{\Phi}_2\boldsymbol{\Gamma}_{12}\boldsymbol{\Phi}_1^{\mathrm{T}}\boldsymbol{\Phi}_1^{*}\boldsymbol{\Gamma}_{12}^{\mathrm{H}}\boldsymbol{\Phi}_2^{\mathrm{H}}\right) = \min\{N, K\}. \tag{6.58}$$

We used $\mathbf{R}_{\mathrm{s}} = \frac{P}{N}\mathbf{I}_N$ since we assume no channel knowledge at source and relays and hence optimal power allocation is not possible, i.e., in every second time slot the source distributes the power P equally among the antennas. The factor $\frac{1}{2}$ is due to the use of two time slots.

In order to evaluate the ergodic capacity performance of this scheme we determine the eigenvalues of the channel covariance matrix

$$\frac{1}{N}\mathbf{R}^{-1}\mathbf{H}\mathbf{H}^{\mathrm{H}} = \left(\boldsymbol{\Phi}_2\boldsymbol{\Gamma}_2\boldsymbol{\Gamma}_2^{\mathrm{H}}\boldsymbol{\Phi}_2^{\mathrm{H}}\sigma_{\mathrm{r}}^2 + \sigma_{\mathrm{d}}^2\mathbf{I}_N\right)^{-1}\boldsymbol{\Phi}_2\boldsymbol{\Gamma}_{12}\boldsymbol{\Phi}_1^{\mathrm{T}}\boldsymbol{\Phi}_1^{*}\boldsymbol{\Gamma}_{12}^{\mathrm{H}}\boldsymbol{\Phi}_2^{\mathrm{H}} \tag{6.59}$$

when the number of antennas N goes to infinity (large-array limit), the antenna separation d remains constant and $\mathbf{H} = \boldsymbol{\Phi}_2\boldsymbol{\Gamma}_{12}\boldsymbol{\Phi}_1^{\mathrm{T}}$. The eigenvalues are asymptotically accurate as $N \to \infty$ and serve as an approximation in the non-asymptotic regime. In the large-array limit we obtain for the eigenvalues:

Theorem 6.3.1. *For $N \to \infty$ the eigenvalues of the channel covariance matrix are given by*

$$\lambda_k\left(\frac{1}{N}\left(\boldsymbol{\Phi}_2\boldsymbol{\Gamma}_2\boldsymbol{\Gamma}_2^{\mathrm{H}}\boldsymbol{\Phi}_2^{\mathrm{H}}\sigma_{\mathrm{r}}^2 + \sigma_{\mathrm{d}}^2\mathbf{I}_N\right)^{-1}\boldsymbol{\Phi}_2\boldsymbol{\Gamma}_{12}\boldsymbol{\Phi}_1^{\mathrm{T}}\boldsymbol{\Phi}_1^{*}\boldsymbol{\Gamma}_{12}^{\mathrm{H}}\boldsymbol{\Phi}_2^{\mathrm{H}}\right) \xrightarrow{N\to\infty} \frac{|h_{1k}|^2}{\sigma_{\mathrm{r}}^2} \tag{6.60}$$

for $k = 1, \ldots, K$ and $h_{2k}, f_k \neq 0\ \forall k$.

Proof of the Theorem. See Appendix 8.

Corollary 6.3.2. *For finite N the eigenvalues of $\frac{1}{N}\mathbf{R}^{-1}\mathbf{H}\mathbf{H}^{H}$ are approximated by*

$$\lambda_k\left(\frac{1}{N}\mathbf{R}^{-1}\mathbf{H}\mathbf{H}^{\mathrm{H}}\right) \approx \frac{N|h_{1k}g_k h_{2k}|^2}{\sigma_{\mathrm{d}}^2 + N|g_k h_{2k}|^2\sigma_{\mathrm{r}}^2} \tag{6.61}$$

for $k = 1, \ldots, K$.

The theorem and corollary stated above have an interesting physical interpretation: From [124] we know that for a uniform linear array with weighting vector $\mathbf{w} = \frac{P}{N}\boldsymbol{\Phi}_k$ at the source the 3-dB beamwidth (half-power points) is $\Delta_{\mathrm{3dB}} = 0.891\frac{\lambda}{Nd}$ and the Rayleigh resolution limit (null-to-null beamwidth) is $\Delta_{00} = 2\frac{\lambda}{Nd}$, i.e., the beams from source to relays become narrower with increasing number of transmit antennas and the spatial overlap between the beams disappear. The same holds for the receive side, if $\mathbf{w} = \boldsymbol{\Theta}_k$ (see Fig. 6.12). In our case the "beams" (eigenvectors of the channel correlation matrix (6.59)) become spatially orthogonal in the large-array limit and the kth eigenvalue of the compound channel matrix depends only on the channel parameters of relay k. Actually, the kth eigenvalue depends only on the relay's first hop channel and the relay noise variance: due to array gain at the destination, the signal-to-noise (SNR) ratio at the destination is determined by the relay noise only.

Asymptotic Ergodic Capacity. We assume here a static relay topology, i.e., the relays does not change their positions during the time of interest. The channel coefficients in the relay-uplink and relay-downlink are random variables that are constant during one block of transmission. With random coding over a large number of independent blocks one can achieve the ergodic capacity of the system.

Combining (6.57), (6.59) and (6.60) we obtain for the asymptotic mutual information:

$$I\left(\mathbf{x}_s; \mathbf{y}_d | \boldsymbol{\Phi}_2 \boldsymbol{\Gamma}_{12} \boldsymbol{\Phi}_1^{\mathsf{T}}, \mathbf{R}\right) = \frac{1}{2} \sum_{k=1}^{K} \log_2 \left(1 + \mathrm{snr}|h_{1k}|^2\right). \tag{6.62}$$

In order to determine the asymptotic ergodic capacity, we assume for the first hop channel coefficients a model that includes path loss and small scale-fading:

$$h_{1k} = \frac{1}{(1 + r_{1k})^{\alpha/2}} x_{1k}. \tag{6.63}$$

Here $1 + r_{1k}$ is the normalized distance between source and relay k, and $x_{1k} \sim \mathcal{CN}(m, \sigma_{\mathrm{h}}^2)$.

Rayleigh Fading. We assume i.i.d. x_{1k} with zero mean $m = 0$, i.e., $|x_{1k}|^2$ has an exponential probability density function and the asymptotic ergodic capacity is obtained by taking the expectation over the channel statistics

$$\begin{aligned}
C^\infty &= \frac{1}{2} \sum_{k=1}^{K} \int_0^\infty \log_2 \left(1 + \frac{\mathrm{snr}|x_{1k}|^2}{(1 + r_{1k})^\alpha}\right) \frac{1}{2\sigma_{\mathrm{h}}^2} e^{-\frac{|x_{1k}|^2}{2\sigma_{\mathrm{h}}^2}} \, \mathrm{d}|x_{1k}|^2 \\
&= \frac{\ln 2}{2} \sum_{k=1}^{K} e^{\frac{1}{\mathrm{snr}_k 2\sigma_{\mathrm{h}}^2}} \mathrm{Ei}\left(-\frac{1}{\mathrm{snr}_k 2\sigma_{\mathrm{h}}^2}\right)
\end{aligned} \tag{6.64}$$

with $\mathrm{snr}_k = \mathrm{snr}/(1 + r_{1k})^\alpha$ and where $\mathrm{Ei}(x)$ is the exponential integral defined as $\mathrm{Ei}(x) = -\int_{-x}^\infty \frac{e^{-t}}{t} \mathrm{d}t$.

Rice Fading. We assume i.i.d. x_{1k} with non-zero mean m per real dimension, i.e., $z = |x_{1k}|^2$ has a non-central chi-square distribution with two degrees of freedom and an upper bound on the

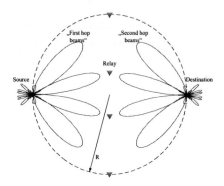

Fig. 6.12: Source and destination with multiple antennas, relays are single antenna nodes. At 17 GHz we have $R = 500\lambda \approx 8.8$m

asymptotic ergodic capacity is then

$$
\begin{aligned}
C_{\text{T3}}^\infty &= \frac{1}{2} \sum_{k=1}^{K} \int_0^\infty \log_2\left(1 + \text{snr}_k z\right) \frac{1}{2\sigma_{\text{h}}^2} e^{-\frac{2m^2+z}{2\sigma_{\text{h}}^2}} J_0\left(\sqrt{2z}\frac{m}{\sigma_{\text{h}}^2}\right) dz \\
&\leq \frac{1}{2} \sum_{k=1}^{K} \int_0^\infty \log_2\left(\text{snr}_k z\right) \frac{1}{2\sigma_{\text{h}}^2} e^{-\frac{2m^2+z}{2\sigma_{\text{h}}^2}} J_0\left(\sqrt{2z}\frac{m}{\sigma_{\text{h}}^2}\right) dz \\
&= \frac{1}{2\ln 2} \sum_{k=1}^{K} \left(\ln\left(2\text{snr}_k m^2\right) - \text{Ei}\left(2m^2\right)\right).
\end{aligned}
\tag{6.65}
$$

where we used in the inequality a high-SNR approximation and in the last equality a result from [125] about the expected-log of a non-central chi-square random variable. $J_n(x)$ denotes the nth-order modified Bessel function of the first kind.

6.3.3 Numerical Results

In this section we present numerical examples in order to demonstrate the accuracy of the asymptotic and approximate eigenvalues given in (6.60) and (6.61) and the ergodic capacities given in (6.64) and (6.65).

Simulation Setup. The setup of the relay network is depicted in Fig. 6.12. As mentioned before we consider here only deterministic relay positions. For random relay topologies, i.e., the relays change their position during the data transmission according to a predefined probability

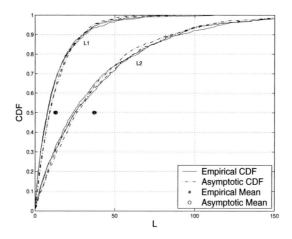

Fig. 6.13: Eigenvalue CDFs of the relay-assisted MIMO Rayleigh channel for $K = 4$ relays and $N = 20$ antennas, where L1 stands for $\lambda_1 = \lambda_4$ and L2 stands for $\lambda_2 = \lambda_3$

distribution, the capacity results have to be averaged over the relay topology. Here the relays are placed such that the angle difference between two relays with respect to source and destination is constant. This is motivated by the observation that by increasing the angle difference (decreasing the relay density) we obtain less spatial crosstalk between the "beams" (eigenvectors of channel correlation matrix) for finite number of antennas and with that the results in (6.60) and (6.61) become more accurate. Note that in the large-array limit the beams become infinitely narrow and no spatial overlap between the beams occurs (channel correlation matrix becomes diagonal).

The first hop channel coefficients are chosen according to (6.63), where in the Rayleigh case we choose $m = 0$ and $\sigma^2 = 1/2$. For Rice channel simulations we choose $m = \sqrt{K_R/2}$, where K_R denotes the Ricean K-factor. The same holds for the second hop channel coefficients. Under the assumption that the relays can measure the receive power the gain coefficients in the amplify-and-forward relays are chosen according to $f_k = \sqrt{P_{rk}/\left(|h_{1k}|^2 P_s + \sigma_r^2\right)}$ where P_{rk} denotes the maximum transmit power of relay k. Note that this is in general a suboptimal power allocation and other strategies can achieve a better performance [41].

Channel Normalization. In order to obtain defined average SNR values at the destination, we normalize the channel matrix for the simulation such, that the average channel gain is equal to the array gain:

$$\widetilde{\mathbf{H}} = \frac{\mathbf{R}^{-1/2}\mathbf{H}}{\sqrt{\mathrm{E}\left\{\|\mathbf{R}^{-1/2}\mathbf{H}\|_{\mathrm{F}}^2\right\}}} N. \tag{6.66}$$

The total average received power is then

$$\mathcal{E}\left\{|\widetilde{\mathbf{H}}\|_{\mathrm{F}}^2\right\} \frac{P_{\mathrm{s}}}{N} = NP_{\mathrm{s}}, \tag{6.67}$$

i.e., it equals the total transmitted power times the receive array gain N. This implies that the eigenvalues have also to be normalized according to

$$\widetilde{\lambda}_k = \frac{\lambda_k\left(\mathbf{R}^{-1}\mathbf{H}\mathbf{H}^{\mathrm{H}}\right)}{\mathcal{E}\left\{\sum_{k=1}^K \lambda_k\left(\mathbf{R}^{-1}\mathbf{H}\mathbf{H}^{\mathrm{H}}\right)\right\}} N^2, \tag{6.68}$$

where we use for $\lambda_k\left(\mathbf{R}^{-1}\mathbf{H}\mathbf{H}^{\mathrm{H}}\right)$ either (6.60) or (6.61). Further we choose an operational frequency of 17GHz , an antenna separation of $d = \lambda/2$ and an average destination SNR of 20dB (averaged over the small-scale fading).

Fig. 6.13 shows the distribution functions (CDFs) of the non-zero eigenvalues of a relay-assisted MIMO system with $K = 4$ relays and $N = 20$ transmit/receive antennas as example. We see that the CDFs based on the asymptotic eigenvalues are quite close to the corresponding empirical distributions.

In Fig. 6.14 we plot capacity vs. number of antennas assuming Rayleigh fading (in relay-uplink and relay-downlink) for different number of relays. We compare the results obtained via the asymptotic (6.60) and the approximated (6.61) eigenvalues with the empirical capacity curve. Capacity scales linearly with number of relays when $K \geq N$, and logarithmic when $K \leq N$ (array gain). Note that the approximative result is very accurate even in the when number of antennas is small.

Fig. 6.15 shows the results for Rice fading. A key observation here is, that the capacity is independent of the Ricean factor (note that the effect of an increased receive power due to the LOS component is removed due to our normalization (6.66)). The rank of the channel matrix and the eigenvalue distribution is determined by the number of relays and their locations. The relays play a role of *active channel shapers* and make the performance of the system insensitive to the small-scale fading statistics.

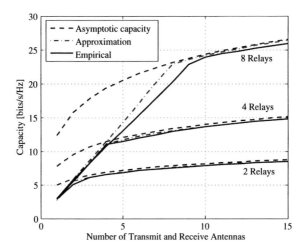

Fig. 6.14: Capacity of the relay-assisted MIMO Rayleigh channel vs. number of antennas N for $K = 2, 4, 8$ relays. Asymptotic capacity refers to (6.64), the approximation refers to (6.61) and empirical means Monte-Carlo simulation of the relay assisted MIMO channel (6.53)

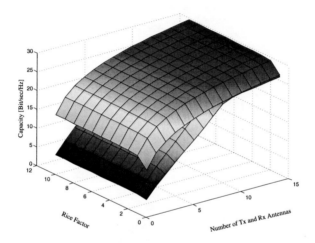

Fig. 6.15: Capacity of the relay-assisted MIMO Rice channel vs. number of antennas N and Rice Factor K_R for 8 relays. The upper surface corresponds to (6.65) and the lower surface is an empirical Monte-Carlo simulation

7 Distributed MIMO Signaling in a Multi-hop Relay Channel

7.1 Introduction

In this chapter we look at a second application of relay networks. We consider a set of wireless nodes where one source-destination pair is assisted by several relays. Each node is equipped wit one antenna and operates a a half-duplex device. As illustrated in Fig.7.1, data from the source is transmitted to the destination via several relays over several hops. For amplify-and-forward relays we propose a relaying protocol that is able to establish a distributed MIMO system between source and destination, although both are not equipped with multiple antennas. We will see that the relays can be used as a virtual antenna array at both side, i.e., source and destination. We compare this system with the case when decode-and-forward relays are used. We show that in many cases the AF scheme is more beneficial and also simpler than the DF scheme.

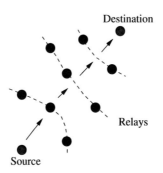

Fig. 7.1: Multi-hop transmission in a multi-relay network with one source-destination pair

7.1.1 Related Work

A similar network setup has been investigated in [126], where a wireless network is considered with one source-destination pair equipped with multiple antennas and several single-antenna relays. The data is transmitted from source to destination via multiple hops. The authors determine the achievable rate for the high-SNR region when amplify-and-forward relays are used and quantify the tradeoff between network size and rate. In [127] resource allocation strategies for multi-hop communication system employing orthogonal frequency-division multiple-access (FDMA) based relaying are investigated. The authors propose bandwidth and power allocations to each relaying hop that maximizes the end-to-end capacity for ergodic frequency-flat Rayleigh channels.

7.1.2 System Model

Transmissions from source to destination takes place sequentially over L hops, whereas in each hop a relay cluster consisting of K relays is involved in the forwarding process. Therefore we consider a wireless network with KL single-antenna nodes: one source node S, one destination node D and $KL - 2$ intermediate relay nodes $\mathcal{R}_k^l$ with $k = 1, \ldots, K$ and $l = 1, \ldots, L - 1$, where $\mathcal{R}_k^l$ is the kth relay in cluster l, cf. Fig. 7.2 and Fig. 7.3. The end-to-end transmission is organized as follows: K codewords are transmitted from source S to the first relay cluster $\{\mathcal{R}_1^1, \ldots, \mathcal{R}_K^1\}$ via orthogonal channels (for example time division multiple access or frequency division multiple access). $L - 2$ time slots each of length T are used for the concurrent transmission of K codewords from the first relay cluster $\{\mathcal{R}_1^1, \ldots, \mathcal{R}_K^1\}$ to the last relay cluster $\{\mathcal{R}_1^{L-1}, \ldots, \mathcal{R}_K^{L-1}\}$ via $L - 2$ hops. Then K orthogonal channels are required to transmit the K codewords from the last relay cluster $\{\mathcal{R}_1^{L-1}, \ldots, \mathcal{R}_K^{L-1}\}$ to the destination node D. Note that in the relay clusters $1 \leq l \leq L - 2$ the signals are jointly transmitted over the same physical channel and cause interference whereas the transmissions in the first and the last hop are interference-free. We will look at two types of forwarding modes: amplify-and-forward (AF) relaying and decode-and-forward (DF) relaying.

Further assumptions: All network nodes are perfectly synchronized and the nodes may not transmit and receive at the same time (half-duplex). The radio range of the nodes does not allow direct communication between source S and destination D, likewise relays $\{\mathcal{R}_1^l, \ldots, \mathcal{R}_K^l\}$ in cluster l only receive signals from relays $\{\mathcal{R}_1^{l-1}, \ldots, \mathcal{R}_K^{l-1}\}$ belonging to cluster $l - 1$.

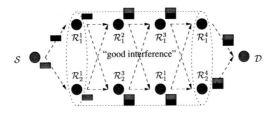

Fig. 7.2: MIMO tunnel: A distributed 2×2 multi-hop MIMO system with amplify-and-forward relays. $K = 2$, $L = 5$.

7.1.3 Signal and Channel Model

Transmissions are organized in time slots of length T with $T = MT_{\text{coh}}$, i.e., M fading realizations are revealed during the transmission of one burst. One time slot contains N symbols of duration T_{s}, i.e., $NT_{\text{s}} = T$. Decoding in the DF case (in the relay nodes as well as in the destination node) is done after the reception of N symbols, i.e., the length of one codeword is T. In the AF case the linear processing (in a relay) is done symbolwise and the decoding in the destination $\mathcal{D}$ after the reception of the codeword (N symbols). The signals transmitted by the source node $\mathcal{S}$ are Gaussian distributed (Gaussian codebook) with an average energy constraint per symbol period [1].

We assume that all channel gains are frequency-flat and time varying. We use a block-fading channel model [121] where a fading coefficient remains constant during a time interval of length T_{coh} (channel coherence time) and changes independently from interval to interval. In this paper we choose all channel gains to be i.i.d. $\mathcal{CN}(0, \nu^2)$ where ν^2 denotes the average channel energy (Rayleigh fading).

In the following sections we consider amplify-and-forward (AF) relaying as well as decode-and-forward (DF) relaying and compare both approaches in terms of information rates (capacity).

7.2 A Distributed MIMO System with Multi-hop Amplify-and-Forward Relays

To illustrate the mode of operation consider Fig. 7.2 as an example. Two codewords are transmitted from $\mathcal{S}$ to $\mathcal{R}_1^1$ and $\mathcal{R}_2^1$ in different time slots (orthogonal channels in the first hop). Intermediate nodes store and forward the received signals simultaneously such that signals received by $\mathcal{R}_1^2$ and $\mathcal{R}_2^2$ are linear superpositions of the signals forwarded by $\mathcal{R}_1^1$ and $\mathcal{R}_2^1$. In the last hop the nodes $\mathcal{R}_1^4$ and $\mathcal{R}_2^4$ cooperate such, that the destination receives the corresponding signals sequentially in time. This establishes a distributed 2×2 MIMO channel between source and destination which we name *MIMO tunnel*.

The symbol vector received by destination $\mathcal{D}$ is given as

$$y_\mathcal{D} = \overbrace{\mathbf{H}_L \prod_{l=1}^{L-1} \mathbf{G}_{L-l} \mathbf{H}_{L-l}}^{\mathbf{H}_\Pi} \mathbf{x}_\mathcal{S} + \underbrace{\sum_{l=1}^{L-1} \prod_{n=0}^{L-1-l} \mathbf{H}_{L-n} \mathbf{G}_{L-n-1} \mathbf{w}_\mathcal{R}^{(l)} + \mathbf{w}_\mathcal{D}}_{\mathbf{n}}, \tag{7.1}$$

where $\mathbf{x}_\mathcal{S} = (x_1, \ldots, x_K)^\mathrm{T}$ is the symbol vector sent by the source $\mathcal{S}$ with i.i.d. $x_k \sim \mathcal{CN}(0, P_s/K)$ and P_s is the average transmit power of the source. $\mathbf{H}_1$ and $\mathbf{H}_L$ are $K \times K$ diagonal channel matrices since the channel from the source to the first relay cluster (entrance to the MIMO tunnel) and the channel from the last relay cluster (exit of the MIMO tunnel) to the destination are orthogonal. $\mathbf{H}_l$ with $l = 2, \ldots, L-1$ are the channel matrices of the intermediate hops, $\mathbf{w}_\mathcal{R}$ and $\mathbf{w}_\mathcal{D}$ denote additive white Gaussian noise at the relays and the destination with variance $\sigma_\mathcal{R}^2$ and $\sigma_\mathcal{D}^2$, respectively. The diagonal matrix $\mathbf{G}_l$ contains the scaling factors of the AF relays with g_k^l the scaling factor of relay $\mathcal{R}_k^l$ that is chosen according to

$$g_k^l = \sqrt{\frac{P_k^l}{\mathcal{E}\left\{\sum_{i=1}^K P_i^{l-1} |h_{k,i}^l|^2\right\} + \sigma_\mathcal{R}^2}} \tag{7.2}$$

where P_k^l is the average transmit power of relay $\mathcal{R}_k^l$ and $h_{k,i}^l$ the channel gain between $\mathcal{R}_i^{l-1}$ and $\mathcal{R}_k^l$. Considering that the channel gains are i.i.d. $\mathcal{CN}(0, \nu^2)$ and setting the average transmit power of the relays equal to P_s/K, i.e., an average transmit power of P_s is used in every channel use, the scaling factor in (7.2) simplifies to

$$g_k^l = g = \sqrt{\frac{P_s/K}{P_s + \sigma_\mathcal{R}^2}}$$

and is the same for all relays. Assuming the destination has perfect channel state knowledge the information rate of the MIMO tunnel measured in bit per tunnel use, i.e., $2K + (L-2)$

consecutive channel uses, is

$$I\left(\mathbf{x}_S; \mathbf{y}_D | \mathbf{H}_0, \ldots, \mathbf{H}_L\right) = \log \det \left(\mathbf{I}_K + \frac{P_s}{K\sigma_D^2} \mathbf{R}^{-1} \mathbf{H}_\Pi \mathbf{H}_\Pi^H\right), \tag{7.3}$$

where the covariance matrix $\sigma_D^2 \mathbf{R}$ of the noise in (7.1) is given by

$$\sigma_D^2 \left(g^2 \sum_{l=1}^{L-1} \prod_{n=0}^{L-1-l} \mathbf{H}_{L-n} \left(\prod_{n=0}^{L-1-l} \mathbf{H}_{L-n}\right)^H \frac{\sigma_R^2}{\sigma_D^2} \mathbf{I}_K + \mathbf{I}_K\right).$$

The ergodic capacity of the MIMO tunnel is obtained by averaging (7.3) over all channel realizations:

$$C_{\text{AF}} = \mathcal{E}\left\{I\left(\mathbf{x}_S; \mathbf{y}_D | \mathbf{H}_0, \ldots, \mathbf{H}_L\right)\right\} \tag{7.4}$$

and may be achieved by random coding over a large number of independent channel realizations [122]. In order to determine the ergodic capacity in (7.4) analytically one has to find the distribution of the eigenvalues of $\mathbf{R}^{-1}\mathcal{H}\mathcal{H}^H$. The eigenvalue distribution of the product channel $\mathcal{H}\mathcal{H}^H$ is given in terms of the Stieltjes transform in [128] and was found by using arguments from random matrix theory. To the best knowledge of the authors the eigenvalue distribution of $\mathbf{R}^{-1}\mathcal{H}\mathcal{H}^H$ is not known so far. First results are available in [78], where Replica methods from statistical physics are applied in order to compute the ergodic capacity of two-hop MIMO relay channels. However, the ergodic capacity of multi-hop MIMO relay channels is still an open problem. Therefore we resort to simulations in order to evaluate the ergodic capacity in (7.4). Numerical examples are given in section 7.4.

7.3 A Distributed MIMO System with Multi-hop Decode-and-Forward Relays

7.3.1 System Model

We compare the MIMO tunnel with a multi-path transmission using decode-and-forward relays where the source sends K independent data streams simultaneously over K network paths, cf. Fig. 7.3. The first stream with rate R_1 is transmitted along path $P_1 : \mathcal{S} \rightarrow \mathcal{R}_1^{(1)} \rightarrow \mathcal{R}_1^{(2)} \rightarrow \cdots \rightarrow \mathcal{R}_1^{(L-1)} \rightarrow \mathcal{D}$, the second stream with rate R_2 along path $P_2 : \mathcal{S} \rightarrow \mathcal{R}_2^{(1)} \rightarrow \mathcal{R}_2^{(2)} \rightarrow \cdots \rightarrow \mathcal{R}_2^{(L-1)} \rightarrow \mathcal{D}$ and so on. Every path consists of L decode-and-forward (DF) wireless single-input

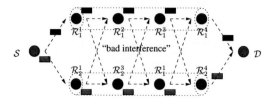

Fig. 7.3: Multi-path transmission (here over two network paths) in a multi-hop wireless network with decode-and-forward relay nodes. $K = 2$, $L = 5$.

single-output (SISO) links. The signal received in path P_k in cluster l ($1 \leq l \leq L-1$) is given as

$$y_k^{(l)} = h_{k,k}^{(l)} x_k^{(l-1)} + \sum_{i=1,i\neq k}^{K} h_{k,i}^{(l)} x_i^{(l-1)} + w_k^{(l)}, \qquad (7.5)$$

where $x_k^{(l-1)} \mathcal{CN}(0, P_s/K)$ is the re-encoded source symbol x_k belonging to path P_k, the sum in (7.5) describes the interference caused by the $K-1$ parallel network paths and $w_k^{(l)}$ is additive white Gaussian noise with variance $\sigma_\mathcal{R}^2$. We refer to orthogonal multi-path transmission when no interference between the network paths occurs (achieved by orthogonal signaling between network paths). Note that the transmissions in the first and the last hop occur over orthogonal channels in any case in order to compare the system with the MIMO tunnel from the previous section.

7.3.2 Achievable Rates

In the following we consider to types of decode-and-forward signaling. By *ergodic signaling* we refer to the case where the codeword length $T = MT_{\text{coh}}$ captures enough channel fluctuations ($M \gg 1$) in order to reveal the ergodic nature the channel such that ergodic capacity is achieved in every link. By *weakest link signaling* the source adapts the rate for a particular path to the weakest channel (link) in that path and the relays belonging to that path decode-and-forward the signals based on this adapted rate. For every scheme we determine the capacity for the case of non-interfering network paths as well as for interfering network paths.

7.3.2.1 Weakest Link Signaling

When the coherence time T_{coh} of the channel gains is large (slow fading) then the overall delay (coding delay plus multi-hop delay) of the source-destination transmission becomes large for ergodic signaling and may be prohibitive for delay-constrained transmissions. However, assuming perfect channel state information at the source, we may choose a variable-rate transmission strategy: rate R_k for path P_k is adapted and matched to the instantaneous channel conditions of path P_k such that the link with the smallest channel gain may support rate R_k. Every relay node in path P_k decodes and re-encodes the symbols based on the code rate R_k. Short codes may be used (codeword length smaller than the coherence time) to allow reliable communication. However, the length of the codewords has still to be long enough to protect the transmissions against noise.

The knowledge of the channel gains at the source may be obtained in a two-way handshake protocol which runs separately for every path: In the first phase training symbols are transmitted along path P_k enabling receiving nodes (intermediate relays and destination) to learn the channel. The channel gains are then propagated back to the source in the second phase. The capacity of one path P_k is determined by the mutual information of the weakest link (the link with deepest channel fade) in path P_k. The average capacity between source and destination for this signaling scheme follows as

$$
C_{\text{WL}} = \sum_{k=1}^{K} \mathcal{E} \left\{ \min_{1 \le l \le L} \left\{ \log \left(1 + \frac{P_{\text{s}}}{K \left(\sigma^2 + P_{\text{i}}^l \right)} |h_{k,k}^l|^2 \right) \right\} \right\}, \tag{7.6}
$$

with $\sigma_D^2 = \sigma_R^2 = \sigma^2$ and where the expectation is taken with respect to the distribution of the channel gains. Further

$$
P_{\text{i}}^l = \begin{cases} \frac{P_{\text{s}}}{K} \sum_{i=1,i\neq k}^{K} |h_{k,i}^l|^2 & ; \ l \in \{2,3,\dots,L-1\} \\ 0 & ; \ l \in \{1,L\} \end{cases} \tag{7.7}
$$

is the interference power caused by the paths used in parallel.

Non-interfering networks paths. Here the parallel network paths do not interfere with each other. This may be achieved by orthogonal signaling (e.g. path separation through different frequency bands). The capacity (7.6) simplifies then to

$$C_{\mathrm{WL}} = K\mathcal{E}\left\{ \min_{1 \leq l \leq L} \left\{ \log\left(1 + \frac{\rho}{K}|h_{1,1}^l|^2\right) \right\} \right\}$$

$$= K\mathcal{E}\left\{ \log\left(1 + \frac{\rho}{K} \min_{1 \leq l \leq L}\left\{|h_{1,1}^l|^2\right\}\right) \right\}$$

$$= K \int\limits_0^\infty \log\left(1 + \frac{\rho}{K}h_{\min}\right) f(h_{\min}) \mathrm{d}h_{\min} \qquad (7.8)$$

with $\rho = P_s/\sigma^2$ and $h_{\min} = \min_{1 \leq l \leq L}\{|h_{1,1}^l|^2\}$. The PDF of $h_{\min}$ is determined in the Appendix as:

$$f(h_{\min}) = \frac{L}{\nu^2}\exp\left(-\frac{Lh_{\min}}{\nu^2}\right). \qquad (7.9)$$

Plugging (7.9) into (7.8) and evaluating the integral yields

$$C_{\mathrm{WL}} = \frac{K}{\ln 2}\exp\left(\frac{KL}{\rho\nu^2}\right) E_1\left(\frac{KL}{\rho\nu^2}\right)$$

$$\leq K\log\left(1 + \frac{\rho\nu^2}{KL}\right) \qquad (7.10)$$

where $E_1(x) = \int_x^\infty \frac{e^{-t}}{t}\mathrm{d}t$. Equation (7.10) follows by applying the inequality $e^x E_1(x) < \ln(1 + 1/x)$ [129] and is obtained also by applying Jensen's inequality to (7.8). Note that for fixed K

$$\lim_{L \to \infty} C_{\mathrm{WL}} = 0,$$

i.e., as the number of hops increases the capacity of the multi-hop transmission drops down even though DF relays are used. The reason is that every additional hop increases the chance to get a channel with a deeper fade than up to now and therefore the probability that the path capacity will be smaller by adding a new hop increases. Therefore, the usual advantage of DF multi-hop systems over AF multi-hop systems due to avoiding noise accumulation is relaxed in a fading multi-hop environment when using a weakest link signaling protocol. For fixed L we obtain for the large-relay limit

$$\lim_{K \to \infty} C_{\mathrm{WL}} = C_{\mathrm{WL}}^\infty \leq \frac{\rho\nu^2 \log e}{L} = C_{\mathrm{WL,u}}^\infty,$$

i.e., capacity saturates for a large number of relays. Increasing the number of parallel network paths allows to transmit additional independent data streams but also lowers the transmit power per relay. The two detrimental effects compensate for each other such that the capacity takes a finite value for large K.

Interfering network paths. Now we turn to the case where the parallel network paths do interfere with each other, i.e., only one orthogonal channel is used for relay transmissions:

$$C_{\text{WL}} = K\mathcal{E}\left\{\min_{1\leq l\leq L}\left\{\log\left(1 + \frac{P_s|h_{1,1}^l|^2}{K\left(\sigma^2 + P_i^l\right)}\right)\right\}\right\}.$$

Since $P_i^l = 0$ for $l \in \{1, L\}$ the minimum channel gain occurs with high probability in one of the intermediate hops $l \in \{2, 3, \ldots, L-1\}$ (where interference occurs), hence we neglect the first and the last hop for the capacity calculation:

$$\begin{aligned} C_{\text{WL}} &\approx K\mathcal{E}\left\{\log\left(1 + \min_{2\leq l\leq L-1}\left\{\frac{|h_{1,1}^l|^2}{\frac{K}{K-1}\sum_{i=2}^{K}|h_{1,i}^l|^2}\right\}\right)\right\} \\ &= K\int_0^\infty \log(1+y)f_Y(y)\mathrm{d}y, \end{aligned} \tag{7.11}$$

where we also assume that $\sigma^2 \ll \frac{P_s}{K-1}\sum_{i=1,i\neq k}^{K}|h_{k,i}^l|^2$ (high SNR approximation). Further we define

$$y = \min_{2\leq l\leq L-1}\left\{\frac{|h_{1,1}^l|^2}{\frac{K}{K-1}\sum_{i=2}^{K}|h_{1,i}^{(l)}|^2}\right\}.$$

The PDF of Y is determined in the Appendix as:

$$f_Y(y) = K(L-2)\left(1 + \frac{K}{K-1}y\right)^{(L-2)(1-K)-1}.$$

The probability $\Pr[Y \leq 1]$ is approximately given by $1 - 2^{-LK}$, i.e., the probability that y takes a small value is very high and we may approximate $\log(1 + y) \approx y\log e$ in (7.11). The capacity follows then by evaluating the integral in (7.11) using this approximation and for $K > 1$ and $L > 2$:

$$C_{\text{WL}} \approx \frac{\log e}{L - \frac{1}{K-1}}.$$

As in the non-interfering case we have for fixed K the limit $\lim_{L\to\infty} C_{\text{WL}} = 0$. For fixed L the large-relay limit becomes

$$\lim_{K\to\infty} C_{\text{WL}} \approx \frac{\log e}{L},$$

i.e., capacity again saturates for a large number of relays.

7.3.2.2 Ergodic Signaling

If we assume that the channel coherence time is small (fast fading) the relays may achieve ergodic capacity of their corresponding links without CSI at the source by choosing codewords long enough in order to sufficiently average out the channel fluctuations [122] but short enough to guarantee an overall delay imposed by multi-hopping and coding that is still feasible. The capacity of path P_k is then determined by the lowest ergodic link capacity in that path. The total ergodic capacity between source and destination follows as

$$C_{\mathrm{erg}} = \sum_{k=1}^{K} \min_{1 \leq l \leq L} \left\{ \mathcal{E} \left\{ \log \left(1 + \frac{P_{\mathrm{s}}}{K \left(\sigma^2 + P_{\mathrm{i}}^l \right)} |h_{k,k}^l|^2 \right) \right\} \right\},$$

where $P_{\mathrm{i}}^{(l)}$ is given in (7.7).

Non-interfering networks paths. The capacity for that scheme follows as

$$\begin{aligned}
C_{\mathrm{erg}} &= K \min_{1 \leq l \leq L} \left\{ \mathcal{E} \left\{ \log \left(1 + \frac{\rho}{K} |h_{1,1}^l|^2 \right) \right\} \right\} \\
&= K \int_0^\infty \log \left(1 + \rho h \right) g(h) \mathrm{d}h \\
&= \frac{K}{\ln 2} \exp \left(\frac{K}{\rho \nu^2} \right) E_1 \left(\frac{K}{\rho \nu^2} \right) \\
&\leq K \log \left(1 + \frac{\rho \nu^2}{K} \right),
\end{aligned}$$

where $h = |h_{1,1}^1|^2$. In contrast to (7.10) the capacity does not degrade with increasing L (the number of hops). Since ergodic signaling at every DF node allows to be robust against deep channel fades the capacity is not determined by the weakest channel link. In the large-relay limit we have

$$\lim_{K \to \infty} C_{\mathrm{erg}} \leq \rho \nu^2 \log e = LC_{\mathrm{WL,u}}^\infty,$$

i.e., with ergodic signaling larger capacity gains are achievable than by using variable-rate coding based on weakest channel links.

Interfering networks paths. The relays use only one orthogonal channel for the transmissions and therefore interfere with each other, i.e., :

$$C_{\text{erg}} \leq K \min_{2 \leq l \leq L-1} \left\{ \mathcal{E} \left\{ \log \left(1 + \frac{P_s |h_{1,1}^l|^2}{K \left(\sigma^2 + P_i^l \right)} \right) \right\} \right\}$$

$$\leq K \mathcal{E} \left\{ \log \left(1 + \frac{|h_{1,1}^l|^2}{\sum_{i=2}^{K} |h_{1,i}^l|^2} \right) \right\}]$$

$$= K \int_0^\infty \log(1+z) f_Z(z) \mathrm{d}z, \tag{7.12}$$

where we again make the assumption that $\sigma^2 \ll \frac{P_s}{K-1} \sum_{\substack{i=1 \\ i \neq k}}^{K} |h_{k,i}^l|^2$ and where

$$z = \frac{|h_{1,1}^l|^2}{\sum_{i=2}^{K} |h_{1,i}^l|^2}.$$

The PDF of Z is given as (see Appendix)

$$f_Z(z) = \frac{K}{\left(1 + \frac{K}{K-1} z \right)^K}. \tag{7.13}$$

The capacity follows by evaluating the integral in (7.12) with the PDF (7.13). For $K/(K-1) \approx 1$ we obtain

$$C_{\text{erg}} \leq \frac{K^2}{K^2 - 2K + 1} \log e.$$

We see that in the limit $\lim_{K \to \infty} C_{\text{erg}} \leq \log e$ the ergodic signaling achieves better capacity results than weakest link signaling. Due to convexity properties of the $\min\{\cdot\}$ operator and by the use of Jensen's inequality it follows that in general:

$$\sum_{k=1}^{K} \mathrm{E} \left[\min_l \left\{ C_{\text{WL},l} \right\} \right] \leq \sum_{k=1}^{K} \min_l \left\{ \mathrm{E} \left[C_{\text{erg},l} \right] \right\} \tag{7.14}$$

$$C_{\text{WL}} \leq C_{\text{erg}}. \tag{7.15}$$

7.4 Numerical Examples

We provide some numerical examples of the capacity performance both for the MIMO tunnel and the multi-path transmission scheme. In Fig.7.4 the curve labeled with *non-interfering DF-MPT* (decode-and-forward multi-path transmission) corresponds to the case where in each layer the average transmit energy per symbol period is equal to P_s and the curve labeled with *non-interfering DF-MPT norm.* corresponds to the case where the capacity is normalized to the number of orthogonal channels K but the average transmit energy per layer is KP_s. We observe that the average capacity $C_{AF} = E[I_{AF}]$ of the MIMO tunnel approaches almost the average capacity of the orthogonal (interference-free) multi-path transmission DF-MPT and clearly outperforms the this scheme for DF-MPT norm. The MIMO tunnel achieves much higher capacity results than DF-MPT when interference between the network routes occurs. The reason for the large degradation of the non-orthogonal multi-path transmission here is our simplified interference model, where each interfering signal is in the average equally strong as the desired signal. Note that the interference from the network paths helps us in the MIMO tunnel to establish a rich scattering environment ("good interference") which is necessary to obtain significant spatial multiplexing gains, whereas the interference in the multi-path transmission approach limits this signaling scheme strongly ("bad interference").

In Fig. 7.5 we compare the two schemes with respect to the number of hops. The capacity decrease of the MIMO tunnel with respect to the number of hops is due to noise accumulation and the fact that a product of Gaussian matrices becomes increasingly ill-conditioned as the number of concatenated matrices grows. In the multi-path transmission approach every additional hop increases the probability of a weaker link in a path and therefore the probability that the path capacity will be smaller by adding a new link (hop) increases.

In Fig. 7.6 we compare the two schemes when ergodic signaling is used for multi-path transmission. The capacity vs. number of relays behaves qualitatively similar as for weakest link signaling with the difference that the rates are higher for ergodic signaling as (7.14) suggests.

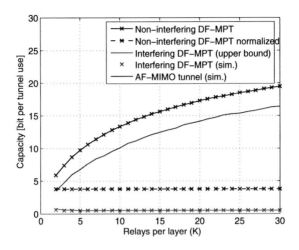

Fig. 7.4: Capacity vs. number of relays per layer for AF MIMO transmission (AF-MIMO) and DF multi-path transmission (DF-MPT) with weakest link signaling, L=5

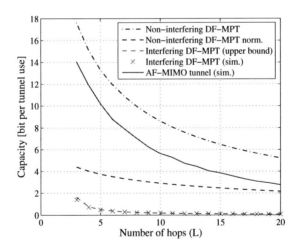

Fig. 7.5: Capacity vs. number of relays per layer for AF MIMO transmission (AF-MIMO) and DF multi-path transmission (DF-MPT) with weakest link signaling, L=5

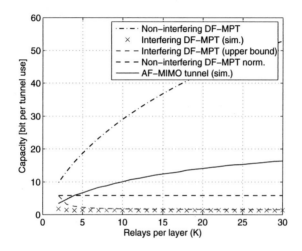

Fig. 7.6: Capacity vs. number of relays per layer for AF MIMO transmission (AF-MIMO) and DF multi-path transmission (DF-MPT) with ergodic signaling, L=5

8 Conclusions and Outlook

In this work we extended the classical relaying protocols to *two-way relay channels* and provided corresponding achievable rate regions. We showed that in an AWGN channel a combined strategy of two-way compress/decode-and-forward achieves the cut-set upper bound on the sum-capacity of the two-way relay channel, when the relay is near to one of the terminals. In a fading two-way relay channel the cut-set upper bound can be achieved by a two-way compress-and-forward strategy.

We proposed *two-way relaying* and *two-path relaying* in order to increase the pre-log factor in half-duplex relay channels. In two-path relaying two relays alternate in transmission and reception. All nodes operate in the same physical channel, thereby causing inter-relay interference. For amplify-and-forward (AF) relays, it was shown that this protocol can recover a significant portion of the half-duplex loss (pre-log factor $\frac{1}{2}$) for weak to moderately strong inter-relay channels. For decode-and-forward (DF) relays we proposed two strategies where the half-duplex loss becomes negligible when the inter-relay channel is very weak or very strong. For the two-way relay protocol, a synchronous bidirectional connection between two half-duplex terminals is established using one half-duplex relay. Hereby, the achievable rate in one direction suffers still from the pre-log factor $\frac{1}{2}$ but since two connections are realized in the same physical channel the sum-rate is considerably higher. We showed that for DF relays the cut-set upper bound is achieved, when the relay is in the proximity of one of the terminals.

Two applications of relay channels have been studied in this thesis. We proposed to use relays for spectral efficient signaling in rank-deficient MIMO channels. These channels fail to provide significant multiplexing gains, since there are not enough spatial degrees of freedom to open up parallel spatial channels. We proposed to use half-duplex amplify-and-forward relays as active scatterer in a way that the eigenvalue structure of the compound two-hop matrix channel improves with respect to the multiplexing capability.

The second application focused on multi-hop relay systems. We proposed two systems, that are capable to achieve a distributed MIMO multiplexing gain with single-antenna nodes. The first system used AF relays and it was shown that the achievable rate decreases with increasing number of hops due to two effects: accumulation of relay noise along the multi-hop chain and singularity of the product channel matrix. The second system used DF relays, where it was shown that for a weakest link signaling protocol, the achievable rate also decreases with increasing number of hops, although there is no noise accumulation along the multi-hop chain. For an ergodic signaling protocol the achievable rate was independent of the number of hops.

There are several open problems for future work:

- For the general two-way relay channel it would be interesting to see how the rate regions improve when the encoders at the terminals are allowed to cooperate via their received symbols. This means that the encoding functions at terminals T_1 and T_2 do not depend on the messages to be transmitted only but also on the past received symbols. The more symbols terminal T_1 receives the more it learns about the message from terminal T_2 and therefore can try to choose its own encoding function based on this additional information. The cooperation could be on top of the cooperation of both terminals with the relay.

- For the spectral efficient half-duplex protocols one possible direction for future work could be to study the outage performance of both protocols, two-way and two-path relaying, for slowly fading channels. From a practical point of view it would be interesting to implement and test both protocols in the RACooN Testbed of the Wireless Communication Group (http://www.nari.ee.ethz.ch/wireless/research/projects/).

- The AF protocol for rank-deficient MIMO channels could be extended to a multi-relay compress-and-forward (CF) strategy. Single antenna relays are not able to decode a MIMO encoded signal (e.g., V-BLAST)[1], however, a CF strategy might work since the relays only quantize their receive signal, they do not try to decode the signals. It would be interesting to compare the CF strategy with the AF strategy proposed in this work. The claim is, that the CF strategy should outperform the AF strategy since it may also provide a rank gain but with a reduced relay noise amplification. Furthermore, the CF strategy may lead to an equivalent channel matrix with a better eigenvalue spread than the AF strategy.

- An interesting open problem is the ergodic capacity of the MIMO tunnel. In order to determine the ergodic capacity analytically one way is to find the distribution of the eigenvalues of the product matrix channel with colored (spatially) noise at the destination. The eigenvalue distribution of the product channel without noise at the relays has been determined in terms of the Stieltjes transform in [128] and was found by using arguments

---

[1]Which means that the AF protocol can not be extended to a DF protocol.

from random matrix theory. The extension of this result to the case of noisy relays is still not found. Another way is the application of Replica methods from statistical physics. First results are obtained in the master's thesis [130], where the ergodic capacity of a two-hop MIMO relay channel with no direct link was found analytically.

Optimal Power Allocation for Two-way Decode-and-forward Relaying

Relay Knows h_1 and h_2

We prove here the optimal power allocation β^* given in (5.37). We assume that the relay T_3 has full knowledge of the backward and forward channels h_1 and h_2. The next section treats the case where the relay has only knowledge of $\mathcal{E}\{|h_1|^2\}$ and $\mathcal{E}\{|h_2|^2\}$. The function to be optimized is the mutual information of the two-way decode-and-forward relaying strategy given in (5.33) and recalled here:

$$I_{\text{sum}}^{\text{DF}} = \max_{\beta} \min\left(I_{\text{MA}}, I_1(\beta) + I_2(1-\beta)\right) \tag{.1}$$

where

$$I_{\text{MA}} = \frac{1}{2} C\left(P_1|h_1|^2 + P_2|h_2|^2\right) \tag{.2}$$

$$I_1(\beta) = \frac{1}{2} \min\left(C\left(P_1|h_1|^2\right), C\left(\beta P_3|h_2|^2\right)\right) \tag{.3}$$

$$I_2(1-\beta) = \frac{1}{2} \min\left(C\left(P_2|h_2|^2\right), C\left((1-\beta)P_3|h_1|^2\right)\right) \tag{.4}$$

and where $C(x) = \log(1+x)$. Note that I_{MA} in (.1) is independent of β and hence we can consider the following maximization instead:

$$\beta^* = \arg\max_{\beta} \left[\min\left\{C\left(P|h_1|^2\right), C\left(\beta P|h_2|^2\right)\right\} + \min\left\{C\left(P|h_2|^2\right), C\left((1-\beta)P|h_1|^2\right)\right\}\right] \tag{.5}$$

where we have chosen $P_1 = P_2 = P_3$ and skipped the factor one-half since it is not important to find the optimal β. For the derivation we look at the cases

i) $|h_1|^2 > |h_2|^2$

ii) $|h_1|^2 \leq |h_2|^2$.

Consider first $|h_1|^2 > |h_2|^2$. Since $C(x)$ is a strictly increasing function it follows that $C\left(P|h_1|^2\right) > C\left(\beta P|h_2|^2\right)$ for all $\beta \in [0,1]$. We therefore have

$$\beta^* = \arg\max_\beta \left[C\left(\beta P|h_2|^2\right) + \min\left\{ C\left(P|h_2|^2\right), C\left((1-\beta)P|h_1|^2\right)\right\}\right]. \tag{.6}$$

Now consider the interval $\beta \in [0, \widetilde{\beta}]$ for which we have $C\left(P|h_2|^2\right) \leq C\left((1-\beta)P|h_1|^2\right)$. In this interval the minimum in (.6) is always given by $C\left(P|h_2|^2\right)$ which is independent of β. Hence if the optimal β falls into the interval $[0, \widetilde{\beta}]$ we have

$$\beta^* = \arg\max_\beta \left[C\left(\beta P|h_2|^2\right) + C P|h_2|^2\right] = \widetilde{\beta}. \tag{.7}$$

The value $\widetilde{\beta}$ is determined from $C\left(P|h_2|^2\right) = C\left((1-\widetilde{\beta})P|h_1|^2\right)$. It follows

$$\widetilde{\beta} = 1 - \frac{|h_2|^2}{|h_1|^2}. \tag{.8}$$

In the interval $\beta \in (\widetilde{\beta}, 1]$ we have $C\left(P|h_2|^2\right) > C\left((1-\beta)P|h_1|^2\right)$ and therefore

$$\beta^* = \arg\max_\beta \left[C\left(\beta P|h_2|^2\right) + C(1-\beta)P|h_1|^2\right]. \tag{.9}$$

Note that (.9) is the sum rate formula of two parallel Gaussian channels with channel gains h_1 and h_2 and the optimal β can be found by applying the Karush-Kuhn-Tucker conditions or solving (.9) by the method of Lagrange. We obtain for β^* in (.9)

$$\beta^* = \frac{1}{2P|h_1|^2} - \frac{1}{2P|h_2|^2} + \frac{1}{2}. \tag{.10}$$

Combining (.7) and (.10) we obtain for $|h_1|^2 > |h_2|^2$

$$\beta^* = \max\left\{ 1 - \frac{|h_2|^2}{|h_1|^2}, \frac{1}{2P|h_1|^2} - \frac{1}{2P|h_2|^2} + \frac{1}{2}\right\}. \tag{.11}$$

Similar considerations can be done when $|h_1|^2 \leq |h_2|^2$. We then obtain

$$\beta^* = \min\left\{ \frac{|h_1|^2}{|h_2|^2}, \frac{1}{2P|h_1|^2} - \frac{1}{2P|h_2|^2} + \frac{1}{2}\right\}. \tag{.12}$$

Proof of Theorem 6.3.1

First we expand (6.59) in a more suitable form

$$\frac{N}{N}\frac{1}{N}\mathbf{R}_{nn}^{-1}\mathbf{H}\mathbf{H}^{\mathrm{H}} = \left(\frac{\boldsymbol{\Phi}_2\boldsymbol{\Gamma}_2\boldsymbol{\Gamma}_2^{\mathrm{H}}\boldsymbol{\Phi}_2^{\mathrm{H}}\sigma_3^2 + \sigma_2^2\mathbf{I}_N}{N}\right)^{-1}\frac{\boldsymbol{\Phi}_2\boldsymbol{\Gamma}_{12}\boldsymbol{\Phi}_1^{\mathrm{T}}\boldsymbol{\Phi}_1^{\star}\boldsymbol{\Gamma}_{12}^{\mathrm{H}}\boldsymbol{\Phi}_2^{\mathrm{H}}}{N^2}. \tag{.13}$$

For $k \neq l$ and $r_{k,l} = \frac{d}{\lambda}\left(\sin\phi_{1k} - \sin\phi_{1l}\right) \notin \mathbb{Z}$ we analyze the product

$$
\begin{aligned}
\boldsymbol{\Phi}_{1k}^{\mathrm{T}}\boldsymbol{\Phi}_{1l}^{\star} &= \sum_{m=0}^{N-1} e^{j2\pi r_{k,l}m} = \sum_{m=0}^{N-1} z_{k,l}^m = \frac{1 - z_{k,l}^N}{1 - z_{k,l}} = \frac{|1 + (-z_{k,l}^N)|}{|1 - z_{k,l}|} \\
&\leq \frac{1 + |z_{k,l}^N|}{|1 - z_{k,l}|} = \frac{1 + |z_{k,l}|^N}{|1 - z_{k,l}|} = \frac{2}{|1 - z_{k,l}|},
\end{aligned} \tag{.14}
$$

where we used the triangle inequality to bound the finite geometric series. For applicability of the bound (.14) we have to demand that the source has to see every relay under a different angle (otherwise the eigenvectors belonging to aligned relays are spatially not separable). Under this separability assumption we have in the large-array limit

$$\frac{1}{N}\boldsymbol{\Phi}_{1k}^{\mathrm{T}}\boldsymbol{\Phi}_{1l}^{\star} \xrightarrow{N\to\infty} 0. \tag{.15}$$

For $k = l$ we have $r_{k,k} = 0$ and $\boldsymbol{\Phi}_{1k}^{\mathrm{T}}\boldsymbol{\Phi}_{1k}^{\star} = N$. Thus we have for the array steering matrix

$$\frac{1}{N}\boldsymbol{\Phi}_1^{\mathrm{T}}\boldsymbol{\Phi}_1^{\star} \xrightarrow{N\to\infty} \mathbf{I}_K. \tag{.16}$$

The same considerations hold for the receive array response matrix

$$\frac{1}{N}\boldsymbol{\Phi}_2^{\mathrm{H}}\boldsymbol{\Phi}_2 \xrightarrow{N\to\infty} \mathbf{I}_K. \tag{.17}$$

Using the property that for a non-singular matrix $\mathbf{X} \in \mathbb{C}^{N\times N}$ we have $\lambda_k(\mathbf{X}^{-1}) = \lambda_k^{-1}(\mathbf{X})$, we can write for $K < N$

$$\lambda_k \left(\left(\boldsymbol{\Phi}_2 \boldsymbol{\Gamma}_2 \boldsymbol{\Gamma}_2^H \boldsymbol{\Phi}_2^H \sigma_3^2 + \sigma_2^2 \mathbf{I}_N \right)^{-1} \right) = \begin{cases} \lambda_k^{-1} \left(\boldsymbol{\Phi}_2^H \boldsymbol{\Phi} \boldsymbol{\Gamma}_2 \boldsymbol{\Gamma}_2^H \sigma_3^2 + \sigma_2^2 \mathbf{I}_K \right) &, k = 1 \dots K \\ 1/\sigma_2^2 &, k = K+1 \dots N \end{cases}$$

(.18)

where we used

$$\lambda_k(c\mathbf{I} + \mathbf{X}) = c + \lambda_k(\mathbf{X}) \tag{.19}$$

$$\text{rk} \left(\boldsymbol{\Phi}_2 \boldsymbol{\Gamma}_2 \boldsymbol{\Gamma}_2^H \boldsymbol{\Phi}_2^H \right) = K \tag{.20}$$

$$\lambda_k(\mathbf{AB}) = \lambda_k(\mathbf{BA}) \tag{.21}$$

for $k = 1, \dots, K$ and where $\mathbf{A} \in \mathbb{C}^{N \times K}$ and $\mathbf{B} \in \mathbb{C}^{K \times N}$. We obtain in the large-array limit for $k = 1, \dots, K$

$$\lambda_k^{-1} \left(\frac{\boldsymbol{\Phi}_2 \boldsymbol{\Gamma}_2 \boldsymbol{\Gamma}_2^H \boldsymbol{\Phi}_2^H \sigma_3^2 + \sigma_2^2 \mathbf{I}_N}{N} \right) \xrightarrow{N \to \infty} \lambda_k^{-1} \left(\boldsymbol{\Gamma}_2 \boldsymbol{\Gamma}_2^H \sigma_3^2 \right) \tag{.22}$$

and

$$\lambda_k \left(\frac{1}{N^2} \boldsymbol{\Phi}_2 \boldsymbol{\Gamma}_{12} \boldsymbol{\Phi}_1^T \boldsymbol{\Phi}_1^\star \boldsymbol{\Gamma}_{12}^H \boldsymbol{\Phi}_2^H \right) \xrightarrow{N \to \infty} \lambda_k \left(\boldsymbol{\Gamma}_{12} \boldsymbol{\Gamma}_{12}^H \right). \tag{.23}$$

The matrices obtained in (.22) and (.23) are diagonal and hence, the asymptotic eigenvalues are the diagonal elements

$$\lambda_k \left(\frac{1}{N} \mathbf{R}_{nn}^{-1} \mathbf{H} \mathbf{H}^H \right) \xrightarrow{N \to \infty} \frac{|h_{1k} g_k h_{2k}|^2}{\sigma_3^2 |g_k h_{2k}|^2} = \frac{|h_{1k}|^2}{\sigma_3^2} \tag{.24}$$

for $k = 1, \dots, K$ when $g_k \neq 0$ and $h_{2k} \neq 0$ (this is satisfied with probability 1 for the channel model given in (6.63)). Obviously, the number of nonzero eigenvalues cannot exceed K since in the asymptotic case we have $\text{rk}(\mathbf{R}_{nn}^{-1} \mathbf{H} \mathbf{H}^H) = \min\{K, N\} = K$, so that $\lambda_k = 0$ for $k = K+1 \dots N$.

Some PDFs used in Chapter 7

Here we determine some PDFs used in Chapter 7.

PDF of $h_{\min}$

Given that $\left\{ h_{1,1}^{(l)} \right\}_{1 \leq l \leq L}$ are i.i.d. $\mathcal{CN}(0, \nu^2)$ it follows that $\left\{ |h_{1,1}^{(l)}|^2 \right\}_{1 \leq l \leq L}$ are i.i.d. according to $\frac{1}{\nu^2} \exp(-\frac{h}{\nu^2})$. From order statistics [131] we know that the minimum of L independent and exponentially distributed random variables with mean ν^2 has again a probability density function (PDF) which is exponential with mean ν^2 / L.

PDF of Y and Z

Define $\left\{ z^{(l)} \right\}_l$ i.i.d. according to $f_Z(z)$ where $z^{(l)} = y_1^{(l)} / y_2^{(l)}$ with $y_1^{(l)} = |h_{1,1}^{(l)}|^2$ and $y_2^{(l)} = \sum_{i=2}^{K} |h_{1,i}|^2$. The PDF of Z is determined by

$$
f_Z(z) = \int_0^\infty y_2 f_{Y_1}(y_2 z) f_{Y_2}(y_2) \mathrm{d}y_2
$$

$$
= \frac{K}{\left(1 + z \frac{K}{K-1} \right)^K}
$$

with

$$
f_{Y_1}(y_2 z) = \frac{1}{\nu^2} \exp(-\frac{y_2 z}{\nu^2})
$$

and

$$f_{Y_2}(y_2) = \frac{\frac{K-1}{K}}{2^{K-1}\Gamma(K-1)\left(\frac{\nu}{\sqrt{(2)}}\right)^{2(K-1)}}$$

$$\times \left(\frac{K-1}{K}y_2\right)^{K-2} \exp\left(\frac{K-1}{K\nu^2}y_2\right)$$

where $\Gamma(n) = (n-1)!$. The PDF of

$$y = \min_{2 \leq l \leq L-1} \left\{ \frac{y_1^{(l)}}{y_2^{(l)}} \right\}$$

is then given by [131]

$$f_Y(y) = (L-2)\left[1 - F_Z(y)\right]^{L-3} f_Z(y)$$
$$= K(L-2)\left(1 + \frac{K}{K-1}y\right)^{(L-2)(1-K)-1}$$

for $K > 1$ and $L > 2$ and where $F_Z(z)$ is the CDF of $f_Z(z)$.

Acronyms

ADC	Analog-to-Digital Converter
AEP	Asymptotic Equipartition Property
AF	Amplify-and-Forward
AoA	Angle of Arrival
AoD	Angle of Departure
AP	Access Point
APEP	Average Pairwise Error Probability
AR	Ad-hoc Relays
AWGN	Additive White Gaussian Noise
BC	Broadcast
BER	Bit Error Ratio
CDF	Cummulative Density Function
CF	Compress-and-Forward
CRC	Cyclic Redundancy Check
CSI	Channel State Information
CSIR	Channel State Information at Receiver
CSIT	Channel State Information at Transmitter
DAC	Digital-to-Analog Converter
dB	Dezibel
DF	Decode-and-Forward
DF-MPT	Decode-and-forward Multi-path Transmission
DMS	Discrete Memoryless Source
DRA	Distributed Relay Array

FD	Frequency Division
FDMA	Frequency Division Multiple Access
FEC	Forward Error Correction
FFT	Fast Fourier Transform
ISM	Industrial Scientific and Medical Band
LO	Local Oscillator
LOS	Line of Sight
MA	Multiple Access
MIMO	Multiple Input Multiple Output
MISO	Multiple Input Single Output
ML	Maximum Likelihood
MMSE	Minimum Mean Square Estimator
MRC	Maximum Ratio Combining
OF	Orthogonalize-and-forward
OFDM	Orthogonal Frequency Division Multiplex
PEP	Pairwise Error Probability
PHY	Physical Layer
RACooN	Radio Access with Cooperative Nodes
SISO	Single Input Single Output
SNR	Signal to Noise Ratio
SVD	Singular Value Decomposition
TD	Time Division
TWC	Two-way Channel
TWRC	Two-way Relay Channel
V-BLAST	Vertical Bell Laboratories Layered Space-Time

WLAN   Wireless Local Area Network

ZF   Zero Forcing

Notation

x	scalar x
$\mathbf{x}$	vector $\mathbf{x}$
$\mathbf{X}$	matrix $\mathbf{X}$
x^n	length-n sequence $x_1, x_2, \ldots, x_n$
$\mathcal{X}$	set of all symbols x
$\mathcal{X}^n$	set of all sequences $x^n = x_1, x_2, \ldots, x_n$ where $x_i \in \mathcal{X}$
$|\mathcal{X}|$	cardinality of the set $\mathcal{X}$
$N(a|x^n)$	number of occurrences of the letter $a \in \mathcal{X}$ in the sequence x^n
P_X	discrete probability distribution of X
P_X^n	discrete probability distribution of X^n
p_X	probability density of X
$\Pr[X \in A]$	probability that $X \in A$
$\mathcal{E}\{X\}$	expectation of X
$\mathcal{V}\{X\}$	variance of X
$Z \sim \mathcal{CN}(m, \sigma^2)$	A circularly symmetric complex Gaussian random variable $Z = X + jY$ where X and Y are i.i.d. normal each with mean m and variance $\frac{\sigma^2}{2}$
$T_\epsilon^n(P_X)$	strongly typical set with respect to P_X and ϵ
$H(X)$	discrete entropy of X
$h(X)$	differential entropy of X
$C(x)$	$\log(1+x)$ or $\mathcal{E}\{\log(1+x)\}$
$I(X;Y)$	mutual information between X and Y

$(\cdot)^{\mathrm{T}}$	transpose
$(\cdot)^*$	conjugate
$(\cdot)^{\mathrm{H}}$	complex conjugate transpose
$\lvert x \rvert$	absolute value of the complex number x
$\mathfrak{Re}\{x\}$	real part of the complex number x
$\mathfrak{Im}\{x\}$	imaginary part of the complex number x
$\det(\mathbf{X})$	determinant of matrix $\mathbf{X}$
$\mathrm{tr}(\mathbf{X})$	trace of matrix $\mathbf{X}$
$\mathrm{rk}(\mathbf{X})$	rank of matrix $\mathbf{X}$
$\lVert \mathbf{X} \rVert_{\mathrm{F}}$	Frobenius norm of matrix $\mathbf{X}$
$\lambda_k(\mathbf{X})$	kth eigenvalue of matrix $\mathbf{X}$
$\mathbf{I}_n$	$n \times n$ identity matrix
$\mathbf{Y} = \mathrm{null}(\mathbf{X})$	columns of $\mathbf{Y}$ span the nullspace of $\mathbf{X}$
$\log(x)$	logarithm of x to the base 2
$\mathrm{mod}(x, y)$	modulus after division, i.e., $\mathrm{mod}(x, y) = x - ny$ where n is the largest integer smaller or equal than x/y
$n!$	factorial of n, i.e., $n! = n(n-1)(n-2)(n-3)\ldots 3 \cdot 2 \cdot 1$

Bibliography

[1] A. Paulraj, R. Nabar, and D. Gore, *Introduction to Space-Time Wireless Communications.* Cambridge University Press, 2003.

[2] T. M. Siep, I. C. Gifford, R. C. Braley, and R. F. Heile, "Paving the way for personal area network standards: an overview of the IEEE P802.15 working group for wireless personal area networks,"

[3] K. J. Negus, A. P. Stephens, and J. Lansford, "HomeRF: wireless networking for the connected home," vol. 7, pp. 20–27, Feb. 2000.

[4] I. Akyildiz, W. Su, Y. Sankarasubramaniam, and E. Cayirci, "A survey on sensor networks ," *IEEE Commun. Mag.*, vol. 40, pp. 102–114, Aug. 2002.

[5] M. Ilyas, *The Handbook of Ad Hoc Wireless Networks.* CRC Press, 2002.

[6] P. Gupta and P. R. Kumar, "The capacity of wireless networks," *IEEE Trans. Inform. Theory*, vol. 46, pp. 388–404, Mar. 2000.

[7] J. H. Winters, "Wireless PHX/LAM system with optimum combining," 1984. United States Patent.

[8] J. Salz, "Digital transmission over cross-coupled linear channels," *Bell Syst. Tech. J.*, vol. 64, July-August 1985.

[9] I. E. Telatar, "Capacity of multi-antenna Gaussian channels," tech. rep., AT&T Bell Laboratories, 1995.

[10] G. J. Foschini, "Layered space-time architecture for wireless communication in a fading environment when using multi-element antennas," *Bell Labs Tech. J.*, vol. 1, pp. 41–59, Autumn 1996.

[11] D. Lenz, B. Rankov, D. Erni, W. Bächtold, and A. Wittneben, "MIMO channel for modal multiplexing in highly overmoded optical waveguides," in *Proc. Proc. IZS*, (Zurich, Switzerland), Feb. 2004.

[12] V. Tarokh, N. Seshadri, and A. Calderbank, "Space-Times Codes for High Data Rate Wireless Communication: Performance Criterion and Code Construction," *IEEE Trans. Inform. Theory*, vol. 44, pp. 744–765, Mar. 1998.

[13] C. Chuah, D. Tse, J. Kahn, and R. Valenzuela, "Capacity scaling in MIMO wireless systems under correlated fading," *IEEE Trans. Inform. Theory*, vol. 48, pp. 637–650, Mar. 2002.

[14] T. M. Cover and A. El Gamal, "Capacity theorems for the relay channel," *IEEE Trans. Inform. Theory*, vol. 25, pp. 572–584, Sept. 1979.

[15] G. Kramer, M. Gastpar, and P. Gupta, "Cooperative strategies and capacity theorems for relay networks," *IEEE Trans. Inform. Theory*, vol. 51, pp. 3037–3063, Sept. 2005.

[16] A. Host-Madsen and J. Zhang, "Capacity bounds and power allocation for wireless relay channels," *IEEE Trans. Inform. Theory*, vol. 51, pp. 2020–2040, June 2005.

[17] M. Khojastepour, A. Sabharwal, and B. Aazhang, "On the capacity of 'cheap' relay networks," in *Proc. Conference on Information Sciences and Systems*, (Princeton), April 2003.

[18] A. Sendonaris, E. Erkip, and B. Aazhang, "User cooperation diversity – Part I & Part II," *IEEE Trans. Commun.*, vol. 51, pp. 1927–1948, Nov. 2003.

[19] N. J. Laneman, "Limiting analysis of outage probabilities for diversity schemes in fading channels," in *IEEE GLOBECOM*, vol. 3, (San Francisco, CA), pp. 1242 – 1246, Dec. 2003.

[20] E. C. van der Meulen, "Three-terminal communication channels," *Adv. Appl. Prob.*, vol. 3, pp. 120–154, 1971.

[21] H. Sato, "Information transmission through a channel with relay," *The Aloha System, University of Hawai, Honolulu, Tech. Rep. B76-7*, pp. 549–552, Mar. 1976.

[22] A. Sendonaris, E. Erkip, and B. Aazhang, "Increasing uplink capacity via user cooperation diversity," in *Proc. IEEE Int. Symposium on Inf. Theory*, p. 156, 1998.

[23] A. Sendonaris, E. Erkip, and B. Aazhang, "User cooperation diversity – Part I & Part II," *IEEE Trans. Commun.*, vol. 51, pp. 1927–1948, Nov. 2003.

[24] N. J. Laneman and G. W. Wornell, "Energy-efficient antenna sharing and relaying for wireless networks," in *Proc. IEEE Wirel. Comm. and Netw. Conf. (WCNC)*, vol. 1, (Chicago, IL), pp. 7 – 12, Sept. 2000.

[25] J. N. Laneman and G. W. Wornell, "Exploiting distributed spatial diversity in wireless networks," in *Proc. Allerton Conf. Comm., Contr. and Comp.*, 2000.

[26] J. N. Laneman, D. N. Tse, and G. W. Wornell, "An efficient protocol for realizing cooperative diversity in wireless networks," in *Proc. IEEE Int. Symposium on Inf. Theory*, p. 294, 2000.

[27] J. N. Laneman, D. N. Tse, and G. W. Wornell, "Cooperative diversity in wireless networks: Efficient protocols and outage behavior," *IEEE Trans. Inform. Theory*, vol. 50, pp. 3062–3080, Dec. 2004.

[28] S. Alamouti, "A simple transmit diversity technique for wireless communications," *IEEE J. Select. Areas Commun.*, vol. 16, pp. 1451–1458, Oct. 1998.

[29] V. Tarokh, H. Jafarkhani, and A. Calderbank, "Space-Times Block Codes from Orthogonal Designs," *IEEE Trans. Inform. Theory*, vol. 45, pp. 1456–1467, July 1999.

[30] A. Wittneben and M. Kuhn, "A new concatenated linear high rate space-time block code," in *Proc. Proc. 55th IEEE Veh. Tech. Conf.*, vol. 1, pp. 289–293, 2002.

[31] M. Kuhn, I. Hammerstroem, and A. Wittneben, "Linear scalable space-time codes: Tradeoff between spatial multiplexing and transmit diversity." submitted to SPAWC, Rome, 2003.

[32] B. X. and J. T. Li, "Decode-amplify-forward: A new class of forwarding strategy for wireless relay channels," in *Proc. Proc. IEEE SPAWC*, (New York City (NY)), June 2005.

[33] A. S. Avestimehr and D. N. C. Tse, "Outage capacity of the fading relay channel in the low SNR regime," *IEEE Trans. Inform. Theory*, 2006. submitted.

[34] J. Boyer, D. D. Falconer, and H. Yanikomeroglu, "A theoretical characterization of the multihop wireless communications channel with diversity," in *Proc. IEEE GLOBECOM*, vol. 2, (San Antonio, TX), pp. 841–845, Nov. 2001.

[35] J. Boyer, D. D. Falconer, and H. Yanikomeroglu, "Multihop diversity in wireless relaying channels," *IEEE Trans. Commun.*, vol. 52, pp. 1820–1830, Oct. 2004.

[36] R. U. Nabar, F. W. Kneubühler, and H. Bölcskei, "Performance limits of amplify-and-forward based fading relay channels," in *IEEE ICASSP*, (Montreal, Canada), May 2004.

[37] R. U. Nabar and H. Bölcskei, "Space-time signal design for fading relay channels," in *Proc. IEEE GLOBECOM*, 2003.

[38] R. U. Nabar, H. Bölcskei, and F. W. Kneubühler, "Fading relay channels: Performance limits and space-time signal design," *IEEE J. Select. Areas Commun.*, vol. 22, pp. 1099–1109, Aug. 2004.

[39] Z. E. Herhold, Patrick and G. Fettweis, "A simple cooperative extension to wireless relaying," in *Proc. 2004 International Zurich Seminar on Communications (IZS)*, (Zürich, Switzerland), pp. 36–39, Feb. 2004.

[40] E. Zimmermann, P. Herhold, and G. Fettweis, "On the performance of cooperative relaying protocols in wireless networks," *European Transactions on Telecommunications*, vol. 16, pp. 5–16, Jan. 2005.

[41] A. Wittneben and B. Rankov, "Impact of cooperative relays on the capacity of rank-deficient MIMO channels," in *Proc. Mobile and Wireless Communications Summit (IST)*, (Aveiro, Portugal), pp. 421–425, 2003.

[42] B. Rankov and A. Wittneben, "On the capacity of relay-assisted wireless MIMO channels," in *Proc. IEEE SPAWC*, (Lisbon, Portugal), July 2004.

[43] T. E. Hunter and A. Nosratinia, "Cooperative diversity through coding," in *Proc. IEEE Int. Symposium on Inf. Theory*, p. 220, 2002.

[44] T. E. Hunter and A. Nosratinia, "Performance analysis of coded cooperation diversity," in *Proc. ICC*, vol. 4, pp. 2688–2692, 2003.

[45] T. E. Hunter and A. Nosratinia, "Distributed protocols for user cooperation in multi-user wireless networks," in *IEEE GLOBECOM*, vol. 6, pp. 3788–3792, 2004.

[46] M. Janani, A. Hedayat, T. E. Hunter, and A. Nosratinia, "Coded cooperation in wireless communications: Space-time transmission and iterative decoding," *IEEE Trans. Signal Processing*, vol. 52, pp. 362–371, Feb. 2004.

[47] T. E. Hunter, S. Sanayei, and A. Nosratinia, "Outage analysis of coded cooperation," *IEEE Trans. Inform. Theory*, vol. 52, pp. 375–391, Feb. 2006.

[48] A. Stefanov and E. Erkip, "Cooperative space-time coding for wireless networks," in *Proc. Inform. Theory Workshop (ITW)*, pp. 50–53, 2003.

[49] A. Stefanov and E. Erkip, "On the performance analysis of cooperative space-time coded systems," pp. 729–734, 2003.

[50] A. Stefanov and E. Erkip, "Cooperative coding for wireless networks," *IEEE Trans. Commun.*, vol. 52, pp. 1470–1476, Sept. 2004.

[51] A. Agustín, E. Calvo, J. Vidal, and O. Muñoz, "Evaluation of turbo coded cooperative retransmission schemes," in *Proc.*, 2004.

[52] A. Agustín, J. Vidal, E. Calvo, M. Lamarca, and O. Muñoz, "Hybrid turbo FEC/ARQ systems and distributed space-time coding for cooperative transmission in the downlink," 2004.

[53] A. Nosratinia, T. E. Hunter, and A. Hedayat, "Cooperative communication in wireless networks," *IEEE Commun. Mag.*, vol. 42, pp. 74–80, Oct. 2004.

[54] J. Boyer, D. D. Falconer, and H. Yanikomeroglu, "A theoretical characterization of the multihop wireless communications channel without diversity," in *Proc. IEEE Int. Symp. PIMRC*, 2001.

[55] M. O. Hasna and M.-S. Alouini, "Performance analysis of two-hop relayed transmissions over rayleigh fading channels," in *Proc. Proc. 56th IEEE Veh. Tech. Conf.*, 2002.

[56] M. O. Hasna and M.-S. Alouini, "End-to-end performance of transmission systems with relays over rayleigh-fading channels," *IEEE Trans. Wireless Commun.*, vol. 2, pp. 1126–1131, Nov. 2003.

[57] M. O. Hasna and M.-S. Alouini, "Harmonic mean and end-to-end performance of transmission systems with relays," *IEEE Trans. Commun.*, Jan. 2004.

[58] M. O. Hasna and M.-S. Alouini, "Optimal power allocation for relayed transmissions over rayleigh-fading channels," *IEEE Trans. Wireless Commun.*, vol. 3, pp. 1999–2004, Nov. 2004.

[59] M. O. Hasna and M.-S. Alouini, "A performance study of dual-hop transmissions with fixed gain relays," *IEEE Trans. Wireless Commun.*, vol. 3, pp. 1963–1968, Nov. 2004.

[60] H. Bölcskei, R. U. Nabar, O. Oyman, and A. J. Paulraj, "Capacity scaling laws in MIMO relay networks," *IEEE Trans. Wireless Commun.*, submitted, April 2004.

[61] B. Schein and R. Gallager, "The gaussian parallel relay network," in *Proc. IEEE Int. Symposium on Inf. Theory*, (Sorrento, (Italy)), p. 22, June 2000.

[62] J. N. Laneman and G. W. Wornell, "Distributed space-time coded protocols for exploiting cooperative diversity in wireless networks," in *IEEE GLOBECOM*, vol. 1, pp. 77–81, 2003.

[63] N. J. Laneman and G. W. Wornell, "Distributed space-time-coded protocols for exploiting cooperative diversity in wireless networks," *IEEE Trans. Inform. Theory*, vol. 49, pp. 2415–2425, Oct. 2003.

[64] P. A. Anghel, G. Leus, and M. Kaveh, "Distributed space-time coding in cooperative networks," in *Proc. IEEE SPAWC*, 2003.

[65] P. A. Anghel, G. Leus, and M. Kaveh, "Multi-user space-time coding in cooperative networks," in *IEEE ICASSP*, vol. 4, pp. IV 73–IV 76, 2003.

[66] P. A. Anghel and M. Kaveh, "On the diversity of cooperative system," in *IEEE ICASSP*, vol. 4, pp. 577–580, 2004.

[67] M. Dohler, F. Said, and H. Aghvami, "Higher order space-time block codes for virtual antenna arrays," pp. 198–203, 2003.

[68] M. Dohler, *Virtual Antenna Arrays*. PhD thesis, King's College London, University of London, 2004.

[69] I. Hammerstroem, M. Kuhn, and A. Wittneben, "Cooperative diversity by relay phase rotations in block fading environments," in *Proc. IEEE SPAWC*, 2004.

[70] Y. Jing and B. Hassibi, "Distributed space-time coding in wireless relay networks- Part I: Basic diversity results," *IEEE Trans. Wireless Commun.*, July 2004. Submitted.

[71] Y. Jing and B. Hassibi, "Distributed space-time coding in wireless relay networks- Part II: Tighter bounds and a more general case," *IEEE Trans. Wireless Commun.*, July 2004. Submitted.

[72] B. Hassibi and B. M. Hochwald, "High-Rate Codes That Are Linear in Space and Time," *IEEE Trans. Inform. Theory*, vol. 48, pp. 1804–1824, July 2002.

[73] H. Sato, "Information Transmission through a Channel with Relay," *The Aloha System, University of Hawai, Honolulu, Tech. Rep. B76-7.*

[74] T. Cover and J. A. Thomas, *Elements of Information Theory.* John Wiley & Sons, 1991.

[75] M. Gastpar, G. Kramer, and P. Gupta, "The multiple-relay channel: Coding and antenna-clustering capacity," in *Proc. IEEE Int. Symposium on Inf. Theory*, (Lausanne, Switzerland), p. 137, June 2002.

[76] A. Host-Madsen, "On the capacity of wireless relaying," in *Proc. 56th IEEE Veh. Tech. Conf.*, Sept. 2002.

[77] M. Khojastepour, B. Aazhang, and A. Sabharwal, "On the capacity of 'cheap' relay networks," in *Proc. Conference on Information Sciences and Systems (CISS)*, (Princeton, NJ), Apr. 2003.

[78] J. Wagner, B. Rankov, and A. Wittneben, "Replica analysis of correlated mimo relay channels," *IEEE Trans. Inform. Theory*, 2006. submitted.

[79] V. Morgenshtern and H. Bölcskei, "Crystallization in large wireless networks," *IEEE Trans. Inform. Theory*, 2006. submitted.

[80] L.-L. Xie and P. R. Kumar, "A network information theory for wireless communication: scaling laws and optimal operation," *IEEE Trans. Inform. Theory*, vol. 50, pp. 748–767, May 2004.

[81] M. Gastpar and M. Vetterli, "On the capacity of wireless networks: The relay case," in *IEEE Infocom*, vol. 3, (New York, NY), pp. 1577–1586, June 2002.

[82] M. Gastpar and M. Vetterli, "On the capacity of large Gaussian relay networks," *IEEE Trans. Inform. Theory*, vol. 51, pp. ?–?, Mar. 2005.

[83] X. Cai and G. B. Yao, Y. Giannakis, "Achievable rates in low-power relay links over fading channels," *IEEE Trans. Commun.*, vol. 53, pp. 184–194, Jan. 2005.

[84] C. T. K. Ng and A. J. Goldsmith, "Capacity and power allocation for transmitter and receiver cooperation in fading channels," in *Proc. ICC*, (Istanbul, (Turkey)), June 2006.

[85] M. Khojastepour, B. Aazhang, and A. Sabharwal, "Bounds on achievable rates for general multi-terminal networks with practical constraints," in *Proc. Information Processing in Sensor Networks*, April 2003.

[86] M. Khojastepour, B. Aazhang, and A. Sabharwal, "Cut-set theorems for multi-state networks," *IEEE Trans. Inform. Theory*, 2004. Submitted.

[87] M. Khojastepour, B. Aazhang, and A. Sabharwal, "Lower bounds on the capacity of gaussian relay channel," in *Proc. Conference on Information Sciences and Systems (CISS)*, (Princeton, NJ), Mar. 2004.

[88] M. Khojastepour, B. Aazhang, and A. Sabharwal, "Improved achievable rates for user cooperation and relay channels," in *Proc. IEEE Int. Symposium on Inf. Theory*, (Chicago, USA), p. 4, June 2004.

[89] K. Azarian, H. El Gamal, and P. Schniter, "On the achievable diversity-multiplexing tradeoff in half-duplex cooperative channels." Submitted to IEEE Trans. Inform. Theory, Apr. 2005.

[90] W. Bo, J. Zhang, and A. Host-Madsen, "On the capacity of mimo relay channels," *IEEE Trans. Inform. Theory*, vol. 51, pp. 29–43, Jan. 2005.

[91] A. Wittneben and B. Rankov, "Distributed antenna systems and linear relaying for gigabit MIMO wireless," in *IEEE Veh. Tech. Conf.*, (Los Angeles, LA), pp. 3624–3630, Sept. 2004.

[92] A. Wittneben and B. Rankov, "MIMO signaling for low rank channels," in *International Symposium on Electromagnetic Theory*, May 2004.

[93] A. Wittneben and B. Rankov, "Distributed antenna arrays versus cooperative linear relaying for broadband indoor MIMO wireless," in *International Conference on Electromagnetics in Advanced Applications*, (Torino, Italy), Sept. 2003.

[94] J. Massey, "Applied digital information theory I." Lecture notes, ETH Zurich, 1997, www.isi.ee.ethz.ch/education/public/pdfs/aditI.pdf.

[95] Y. Oohama, "Gaussian multiterminal source coding," *IEEE Trans. Inform. Theory*, vol. 43, pp. 1912–1923, Nov. 1997.

[96] I. Csiszar and J. Körner, *Information Theory: Coding Theorems for Discrete Memoryless Systems*. Academic Press, 1981.

[97] G. Kramer, "Mulitple user information theory." Mini Course on Selected Topics in Information Theory, ETH Zurich, 2006.

[98] A. Wyner and J. Ziv, "The rate-distortion function for source coding with side information at the decoder," *IEEE Trans. Inform. Theory*, vol. 22, pp. 1–10, Jan. 1976.

[99] G. Kramer, "Models and theory for relay channels with receive constraints," in *Proc. Allerton Conf. Comm., Contr. and Comp.*, (Monticello, IL), 2004.

[100] A. El Gamal and N. Hassanpour, "Capacity theorems for the relay-without-delay channel," *Proc. Allerton Conf. Comm., Contr. and Comp.*, Sept. 2005.

[101] A. Lapidoth and G. Kramer, "Topics in multi-terminal information theory: The relay channel." Lecture Notes, ETH Zurich.

[102] M. Katz and S. Shamai (Shitz), "Relaying protocols for two co-located users," *IEEE Trans. Inform. Theory*, 2006. Submitted.

[103] D. Tse and P. Viswanath, *Fundamentals of Wireless Communication*. Cambridge University Press. To be published.

[104] R. U. Nabar, H. Bölcskei, and F. W. Kneubühler, "Fading relay channels: Performance limits and space-time signal design," *IEEE J. Select. Areas Commun.*, vol. 22, pp. 1099–1109, Aug. 2004.

[105] C. E. Shannon, "Two-way communication channels," in *Proc. 4th Berkeley Symp. Math. Stat. and Prob.*, vol. 1, pp. 611–644, 1961.

[106] T. S. Han, "A general coding scheme for the two-way channel," *IEEE Trans. Inform. Theory*, vol. 30, pp. 35–44, Jan. 1984.

[107] A. Host-Madsen, "Capacity bounds for cooperative diversity," *IEEE Trans. Inform. Theory*, vol. 52, pp. 1522–1544, Apr. 2006.

[108] B. Rankov and A. Wittneben, "Achievable rate regions for the two-way relay channel," *IEEE Trans. Inform. Theory*. in preparation.

[109] H. Bölcskei, R. U. Nabar, O. Oyman, and A. J. Paulraj, "Capacity scaling laws in MIMO relay networks," *IEEE Trans. Wireless Commun.*, 2006. To appear.

[110] N. J. Laneman, D. N. Tse, and G. W. Wornell, "Cooperative diversity in wireless networks: Efficient protocols and outage behavior," *IEEE Trans. Inform. Theory*, vol. 50, pp. 3062–3080, Dec. 2004.

[111] O. Munoz, A. Augustin, and J. Vidal, "Cellular capacity gains of cooperative MIMO transmission in the downlink," in *Proc. IZS*, (Zurich, Switzerland), pp. 22–26, Feb. 2004.

[112] H. Hu, H. Yanikomeroglu, D. D. Falconer, and S. Periyalwar, "Range extension without capacity penalty in cellular networks with digital fixed relays," in *IEEE GLOBECOM*, vol. 5, (Dallas, (TX)), pp. 3053–3057, Nov. 2004.

[113] T. J. Oechtering and A. Sezgin, "A new cooperative transmission scheme using the space-time delay code," in *Proc. ITG Workshop on Smart Antennas*, (Munich, Germany), pp. 41–48, Mar. 2004.

[114] A. Wittneben, "A new bandwidth efficient transmit antenna modulation diversity scheme for linear digital modulation," in *Proc. ICC*, vol. 3, pp. 1630–1634, 1993.

[115] A. Ribeiro, X. Cai, and G. B. Giannakis, "Opportunistic multipath for bandwidth-effcicent cooperative networking," in *IEEE ICASSP*, vol. 4, (Montreal, (CA)), pp. 549–552, May 2004.

[116] Y. Fan and J. Thompson, "Recovering multiplexing loss through successive relaying," *IEEE Trans. Wireless Commun.*, June 2006. submitted.

[117] B. Rankov and A. Wittneben, "Spectral efficient protocols for nonregenerative half-duplex relaying," in *Proc. Allerton Conf. Comm., Contr. and Comp.*, (Monticello, IL), Oct. 2005.

[118] M. Kuhn, S. Berger, I. Hammerström, and A. Wittneben, "Power-line enhanced cooperative wireless communications," *IEEE J. Select. Areas Commun.*, 2006. To appear.

[119] D. Gesbert, H. Boelcskei, D. Gore, and A. Paulraj, "Outdoor MIMO Wireless Channels: Models and Performance Prediction," *IEEE Trans. Commun.*, vol. 50, Dec. 2002.

[120] D. Chizhik, G. J. Foschini, M. J. Gans, and R. Valenzuela, "Keyholes, correlations, and capacities of multielement transmit and receive antennas," *IEEE Trans. Commun.*, vol. 1, pp. 361–368, Apr. 2002.

[121] L. Ozarow, S. Shamai (Shitz), and A. Wyner, "Information theoretic considerations for cellular mobile radio," *IEEE Trans. Veh. Technol.*, vol. 43, pp. 359–378, May 1994.

[122] E. Biglieri, J. Proakis, and S. Shamai, "Fading channels: Information-theoretic and communications apects," *IEEE Trans. Inform. Theory*, vol. 44, pp. 2619–2692, Oct. 1998.

[123] O. Munoz, J. Vidal, and A. Augustin, "Non-regenerative MIMO relaying with channel state information," in *IEEE ICASSP*, (Philadelphia, PA), pp. 361–364, Mar. 2005.

[124] H. L. Van Trees, *Optimum Array Processing: Part IV of Detection, Estimation and Modulation Theory.* John Wiley & Sons, 2002.

[125] A. Lapidoth and S. M. Moser, "Capacity bounds via duality with applications to multiple-antenna systems on flat fading channels," *IEEE Trans. Inform. Theory*, vol. 49, pp. 2426–2467, Oct. 2003.

[126] S. Borade, L. Zheng, and R. Gallager, "Maximizing Degrees of Freedom in Wireless Networks," in *Proc. Proc. Allerton Conf. Comm., Contr. and Comp.*, Oct. 2003.

[127] M. Dohler, A. Gkelias, and H. Aghvami, "A Resource Allocation Strategy for Distributed MIMO Multi-Hop Communication Systems," *IEEE Commun. Lett.*, vol. 8, pp. 99–101, Feb. 2004.

[128] R. Müller, "On the asymptotic eigenvalue distribution of concatenated vector-valued fading channels," *IEEE Trans. Inform. Theory*, vol. 48, pp. 2086–2091, July 2002.

[129] M. Abramovitz and I. Stegun, *Handbook of Mathematical Functions.* Dover Publications, Inc., New York, 1972.

[130] J. Wagner, "Replica analysis of correlated MIMO relay channels." Master's Thesis at the Communication Technology Laboratory, ETH Zurich. Supervised by Boris Rankov and Armin Wittneben.

[131] A. Papoulis and S. U. Pillai, *Probability, Random Variables and Stochastic Processes.* McGraw-Hill, 4th ed., 2002.

Curriculum Vitae

Name: **Boris Rankov**
Birthday: April 23, 1974
Birthplace: Dietikon ZH, Switzerland

Education

08/2002 –   **ETH Zurich**
11/2006   PhD studies in Electrical Engineering, Communication Technology Laboratory.
            Degree as Dr. sc. ETH Zurich.

10/1998 –   **ETH Zurich**
04/2002   Studies in Electrical Engineering. Focus on Communications, Information Theory
            and Digital Signal Processing. Degree as Dipl. El. Ing. ETH.

11/1994 –   **University of Applied Sciences Winterthur (ZHW)**
11/1998   Studies in Electrical Engineering. Focus on Signal Processing with application to
            Control and Communication Theory. Degree as Dipl. El. Ing HTL.

08/1990 –   **Sulzer AG**
07/1994   Apprenticeship as tracer ("Maschinenzeichner").

Experience

08/2002 –   **Communication Technology Laboratory ETH Zurich**
11/2006   Research and teaching assistant in wireless communication theory.

04/2003 and   **University of Applied Sciences Winterthur (ZHW)**
04/2004   Guest lecturer on the topic of System Identification.

02/2000 –   **IFA – The Knowledge Company**
03/2004   Lecturer for Financial Mathematics, Statistics, Algebra, Communication Networks.

04/2001 –   **Predict AG**
07/2001   Explorative data analysis of a large customer data base of UBS using statistical
                data mining tools.

10/2000 –   **Technical College Winterthur (Technikerschule)**
07/2001   Lecturer for Mathematics (Analysis), Electrical Engineering.

08/2000 –   **Ascom Systec AG**
10/2000   Development and implementation of algorithms for a powerline modem.

08/1998 –   **Sulzer Innotec AG**
10/1998   Development and implementation of algorithms for signal analysis of rotary machines.

Publications

1. B. Rankov and A. Wittneben, **Spectral Efficient Protocols for Half-duplex Fading Relay Channels**, IEEE Journal on Selected Areas in Communications, Feb. 2007, to appear.

2. B. Rankov and A. Wittneben, **Distributed Antenna Systems versus Cooperative Relaying for Broadband Indoor MIMO Systems**, *Distributed Antenna Systems: Open Architecture for Future Wireless Communications*, CRC Press, 2006, to appear as book chapter.

3. B. Rankov and A. Wittneben, **Achievable Rate Regions for the Two-way Relay Channel**, IEEE Int. Symposium on Information Theory, Seattle, USA, July 2006.

4. B. Rankov and A. Wittneben, **Spectral Efficient Signaling for Half-duplex Relay Channels**, Asilomar Conference on Signals, Systems, and Computers 2005, Pacific Grove, USA, 2005.

5. B. Rankov and A. Wittneben, **Spectral Efficient Protocols for Nonregenerative Half-duplex Relaying**, Allerton Conference on Communication, Control, and Computing, Monticello, USA, 2005.

6. E. Auger, B. Rankov, M. Kuhn and A. Wittneben, **Time Domain Precoding for MIMO-OFDM Systems**, International OFDM-Workshop, Hamburg, 2005.

7. B. Rankov and A. Wittneben, **Distributed Spatial Multiplexing in a Wireless Network**, Asilomar Conference on Signals, Systems, and Computers, Pacific Grove, USA, 2004.

8. A. Wittneben and B. Rankov, **Distributed Antenna Systems and Linear Relaying for Gigabit MIMO Wireless**, IEEE Vehicular Technology Conference, VTC Fall 2004, Los Angeles, USA, 2004.

9. B. Rankov and A. Wittneben, **On the Capacity of Relay-Assisted Wireless MIMO Channels**, Signal Processing Advances in Wireless Communications, Lissabon, Portugal, 2004.

10. A. Wittneben and B. Rankov, **MIMO Signaling for Low Rank Channels**, International Symposium on Electromagnetic Theory, 2004.

11. D. Lenz, B. Rankov, D. Erni, W. Bächtold and A. Wittneben, **Modal Multiplexing in Highly Overmoded Optical Waveguides**, Progr. in Electromagne. Research Symp, 2004.

12. D. Lenz, B. Rankov, D. Erni, W. Bächtold and A. Wittneben, **MIMO Channel for Modal Multiplexing in Highly Overmoded Optical Waveguides**, International Zurich Seminar on Communications, IZS, 2004.

13. I. Hammerström, M. Kuhn, B. Rankov and A. Wittneben, **Space-Time Processing for Cooperative Relay Networks**, IEEE Vehicular Technology Conference, Orlando, USA, 2003.

14. A. Wittneben and B. Rankov, **Distributed Antenna Arrays versus Cooperative Linear Relaying for Broadband Indoor MIMO Wireless**, International Conference on Electromagnetics in Advanced Applications, Torino, Italy, 2003.

15. A. Wittneben and B. Rankov, **Impact of Cooperative Relays on the Capacity of Rank-Deficient MIMO Channels**, IST Summit on Mobile and Wireless Communications, Aveiro, Portugal, 2003.

16. J. Milek, F. Kraus, D. Lenz and B. Rankov, **Model-based Monitoring Of Huge Financial Databases**, Triennial World Congress, Barcelona, Spain, 2002.